Recent Climate Change Impacts on Mountain Glaciers

Wiley-Blackwell Cryosphere Science Series

Permafrost, sea ice, snow, and ice masses ranging from continental ice sheets to mountain glaciers, are key components of the global environment, intersecting both physical and human systems. The study of the cryosphere is central to issues such as global climate change, regional water resources, and sea level change, and is at the forefront of research across a wide spectrum of disciplines, including glaciology, climatology, geology, environmental science, geography and planning.

The Wiley-Blackwell Cryosphere Science Series comprises volumes that are at the cutting edge of new research, or provide a focused interdisciplinary reviews of key aspects of the science.

Series Editor

Peter G Knight, Senior Lecturer in Geography, Keele University

Recent Climate Change Impacts on Mountain Glaciers

Mauri Pelto

Nichols College
USA

WILEY Blackwell

Library of Congress Cataloging-in-Publication data has been applied for:

[9781119068112]

A catalogue record for this book is available from the British Library.

Wiley also publishes its books in a variety of electronic formats. Some content that appears in print may not be available in electronic books.

Cover image: © Jill Pelto

Set in 10/12pt, WarnockPro by SPi Global, Chennai, India
Printed and bound in Singapore by Markono Print Media Pte Ltd

10 9 8 7 6 5 4 3 2 1

Contents

Series Preface

Permafrost, sea ice, snow, and ice masses ranging from continental ice sheets to mountain glaciers are key components of the global environment, intersecting both physical and human systems. The scientific study of the cryosphere is central to issues such as global climate change, regional water resources, and sea-level change. The cryosphere is at the forefront of research across a wide spectrum of disciplinary interests, including glaciology, climatology, geology, environmental science, geography, and planning.

The Wiley-Blackwell Cryosphere Science series serves as a framework for the publication of specialist volumes that are at the cutting edge of new research or provides a benchmark statement in aspects of cryosphere science, where readers from a range of disciplines require a short, focused, state-of-the-art text. These books lie at the boundary between research monographs and advanced textbooks, contributing to the development of the discipline, incorporating new approaches and ideas, and providing a summary of the current state of knowledge in tightly focused topic areas. The books in this series are, therefore, intended to be suitable both as case studies for advanced undergraduates and as specialist texts for postgraduate students, researchers, and professionals.

Cryosphere science is in a period of rapid development, driven in part by an increasing urgency in our efforts to understand the global environmental system and the ways in which human activity impacts it. This rapid development is marked by the emergence of new techniques, concepts, approaches, and attitudes.

Recent Climate Change Impacts on Mountain Glaciers is an exciting addition to the series. The response of mountain glaciers to changing climate in the last few decades has been hugely significant in many locations of the world. This book addresses issues that are important both within cryosphere science and more broadly in the context of global concerns about the challenges and impacts of environmental change.

Peter G Knight
October 2015

Foreword

The goal of this volume is to tell the story, glacier by glacier, of response to climate change from 1984 to 2015. Of the 165 glaciers examined in 10 different alpine regions, 162 have retreated significantly. Glaciers are considered by the IPCC as a star climate indicator, and to people they are an iconic example of climate change. From 1981 to 2015 I have spent every summer working on alpine glaciers examining how they respond. It is evident that the changes are significant, not happening at a "glacial" pace, and are profoundly affecting alpine regions. The World Glacier Monitoring Service (2008) provided a brief examination of glacier change from region to region around the world. There is a consistent result that reverberates from mountain range to mountain range, emphasizing that though regional glacier and climate feedbacks differ, global changes are driving the response. In this book we look at 10 different glaciated regions around the world, and the individual glaciers in each, offering a different tune to the same chorus of glacier volume loss in the face of climate change (Fig. 1).

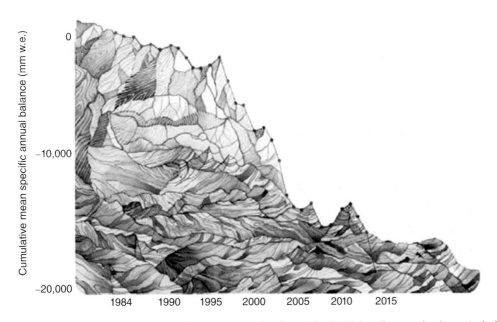

Figure 1 Decrease in glacier mass balance based on data from Pelto (2015a) to illustrate the dramatic decline in North Cascade glaciers, WA. (Watercolor, Jill Pelto 2015.)

1

Alpine Glaciers: An Introduction

1.1   Glacier Observation Programs

Glaciers have been studied as sensitive indicators of climate for more than a century (Forel, 1895; Zemp *et al.*, 2015). Glacier fluctuations in terminus position, mass balance, and area are recognized as one of the most reliable indicators of climate change (Haeberli, Cihlar, and Barry, 2000; Oerlemans, 2005). The recognition of glacier sensitivity to climate led to the development of a global reporting system for glacier terminus change and glacier mass balance during the International Geophysical Year (IGY). Today, this system is managed by the World Glacier Monitoring Service (WGMS). WGMS annually collects standardized observations on changes in mass, volume, area, and length of glaciers with time. This data on individual glacier fluctuations has been enhanced and supplemented in recent years by glacier inventories derived from satellite imagery. Glacier fluctuation and inventory data are today high-priority key variables in climate system monitoring (Sharp *et al.*, 2015; Pelto, 2015b) (Fig. 1.1).

Observations of alpine glaciers most commonly focus on changes in terminus behavior, to identify glacier response to climate changes (Forel, 1895). A number of nations have long-running annual terminus survey programs: Austria, Italy, Switzerland, Norway, and Iceland (WGMS, 2012). The data set of terminus change compiled by the WGMS has 42,000 measurements on 2000 glaciers (Zemp *et al.*, 2015).

Annual mass balance measurements are the most accurate indicator of short-term glacier response to climate change (Haeberli, Cihlar, and Barry, 2000; Zemp, Hoelzle, and Haeberli, 2009). Annual mass balance is the change in mass of a glacier during a year resulting from the difference between net accumulation and net ablation. The importance of monitoring glacier mass balance was recognized during the IGY in 1957. For the IGY, a number of benchmark glaciers around the world were chosen where mass balance would be monitored. This network continued by the WGMS has proven valuable with a total of annual glacier observations; from 1985 to 2014, the average number of glaciers reporting annual mass balance has been approximately 100. Thirty-seven of these are considered reference glaciers with at least a continuous 30-year record of mass balance.

In addition, the advent of frequent high-resolution satellite imagery has allowed for the completion of global mountain glacier inventories led by the Global Land Ice Measurements from Space (GLIMS) and the Randolph Glacier Inventory (RGI) (Arendt *et al.*, 2012; Pfeffer *et al.*, 2014). Detailed repeated inventories have developed a standard approach and also identified changes through time (Kääb *et al.*, 2002; Fischer *et al.*, 2014; Radić and Hock, 2014). The inventories focus on using standard methodologies to define glacier outlines and glacier attributes. Typical attributes include area, length, slope, aspect, terminal environment, elevation range, and shape classification. Satellite images can also be used to map transient snowlines (TSLs), the snowline separating the ablation and accumulation zone

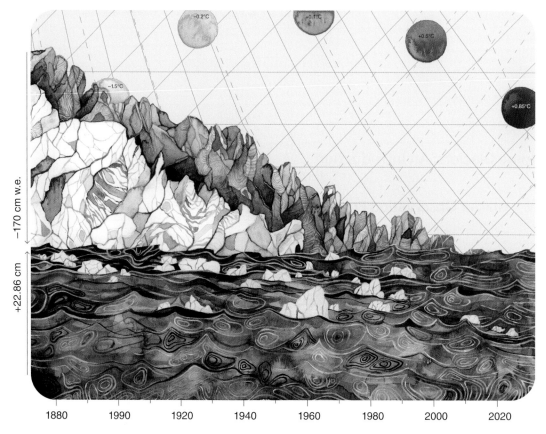

Figure 1.1 Decrease in Glacier Mass Balance based on data from Pelto (2015a) to illustrate the dramatic decline in North Cascades glaciers, WA. (Watercolor, Jill Pelto 2015).

during the summer; the end of summer TSL represents the equilibrium line altitude (ELA) on most alpine glaciers that lack superimposed ice formation (Østrem, 1975; Mernild *et al.,* 2013). These data provide baseline information for an assessment of glacier changes.

The geodetic inventories assess glacier area and in many cases glacier volume. ICESat and other instruments provide elevation change data to compliment areal extent change assessment (Neckel *et al.,* 2014). The remote sensing geodetic inventories and the field glaciological observations both indicate that rates of early twenty-first-century mass loss are historically unprecedented at global scale (Zemp *et al.,* 2015). The largest negative mass balances have occurred in one of the last two decades, depending on the region (WGMS, 2013). The decadal mean annual mass balance was −221 mm in the 1980s, −389 mm in the 1990s, and −726 mm for 2000s. The continued large negative annual balances reported indicate that glaciers are not approaching equilibrium (Pelto, 2010). The strong negative mass balance suggests that glaciers of many regions are committed to further volume loss even under current climatic conditions (Zemp *et al.,* 2015). Radić and Hock, (2014) indicate that future climate change will enhance the mass losses substantially.

The RGI version 3.2 was completed in 2014, compiling digital outlines of glaciers, excluding the ice sheets using satellite imageries from 1999 to 2010. The inventory identified 198,000 glaciers, with a total extent estimated at $726,800 \pm 34,000 \, km^2$ (Pfeffer *et al.,* 2014). An earlier RGI 2.0 has been used to estimate global alpine glacier volume at ~150,000 Gt (Radić and Hock, 2014), quantifying the important role as a water resource and potential contributor to sea-level rise.

This information on glacier mass balance and terminus change has been collected and made available from internationally coordinated efforts (WGMS, 2011, 2013). GLIMS and the RGI have made available their glacier inventory data as well (Arendt *et al.*, 2012; Pfeffer *et al.*, 2014). This is a wealth of information on the state of glaciers.

1.2   Importance of Mountain Glaciers

Mountain glaciers are important as water resources for agriculture, hydropower, aquatic life, and basic water supply (Schaner *et al.*, 2012; Bliss, Hock, and Radić, 2014). Alpine glaciers in many areas of the world are important for water resources – melting in the summer when precipitation is lowest and water demand from society is largest. The timing and magnitude of glacier melt are sensitive to climate change; hence, rational water resource management depends on understanding future changes in water resources from glaciated mountain ranges (Immerzeel, Beek van, and Bierkens, 2010).

Mountain glaciers have also contributed to sea-level rise (Radić *et al.*, 2013; Marzeion, Jarosch, and Hofer, 2012). The annual contribution has been approximately 1 mm a^{-1} during the twentieth century since (Marzeion, Jarosch, and Hofer, 2012). Mountain glaciers can also increase local natural hazards such as glacial lake outburst floods (Bajracharya and Mool, 2009).

1.3   Glacier Terminus Response to Climate Change

In this book, we examine glacier responses during the 1985–2015 period, with the primary climate change being the global temperature rise since 1976 (GISTEMP Team, 2015). Changes in mass balance control a glacier's long-term behavior. Terminus and glacier area changes are then impacted with a lag time for both an initial and more complete response.

For any glacier, there is a lag time (*Ts*) between a significant climate change and the initial observed terminus response (Paterson, 1994); this is also referred to as the reaction time of the glacier. It should be noted that *Ts* cannot be considered a physical property of a glacier and is expected to depend on the mass balance history and physical characteristics of the glacier.

In addition, for each glacier there is a response time to approach a new steady state for a given climate-driven mass balance change (*Tm*), referred to as length of memory by Johannesson, Raymond, and Waddington (1989). They defined *Tm* as the timescale for exponential asymptotic approach to a final steady state (approximately 63% of a full response), resulting from a sudden change in climate to a new constant climate. The magnitudes of *Ts* and *Tm* are crucial to interpreting past and current glacier fluctuations and predicting future changes (Paterson, 1994; Johannesson, Raymond, and Waddington, 1989).

For glaciers in the North Cascades, Washington, Pelto and Hedlund (2001) found a *Ts* of 10–20 years and a *Tm* of 20–100 years.

1.3.1   Equilibrium Response

Typically, glacier terminus retreat results in the loss of the lowest elevation region of the glacier. Since higher elevations are cooler than lower elevations, the disappearance of the lowest portion of the glacier reduces overall ablation, thereby increasing mass balance and potentially reestablishing equilibrium (Pelto, 2010). Typically, a glacier's thinning is greatest at the terminus, and at some distance above the terminus; usually in the accumulation zone, the glacier is no longer thinning appreciably even during retreat (Schwitter and Raymond, 1993). This behavior of greatest thinning at the terminus and limited thinning in the accumulation zone suggests a glacier that will retreat to a new stable position (Schwitter and Raymond, 1993).

A period of sustained positive mass balance will lead to an increase in glacier thickness, an increase in velocity, and eventually an advance. The advance expands the area of the glacier at the lowest elevations where mass balance is more negative. When the expansion at low elevation is sufficient to offset the increased mass balance, the retreat will end as equilibrium is reached.

1.3.2 Disequilibrium Response

In recent years, an increasing number of glaciers have been identified to be experiencing a disequilibrium response to climate (Pelto, 2010; Carturan *et al.*, 2013). There is no point to which such a glacier can retreat to reach equilibrium and the glacier would then disappear. For alpine glaciers, typically 50–70% of the glacier must retain snow cover even at the conclusion of the melt season to be in equilibrium, this is referred to as the accumulation area ratio (AAR). Without a substantial consistent accumulation area, a glacier cannot survive. If a nonsurging alpine glacier is experiencing extensive thinning and marginal retreat in the accumulation zone of the glacier, it lacks a persistent accumulation (Pelto, 2010). The result is a more unstable form of retreat with substantial thinning throughout the length and breadth of the glacier. A glacier in this condition is unlikely to be able to survive in anything similar to its present extent given the current climate. This is evident in satellite or aerial photographs of glaciers. The emergence of bedrock outcrops or the recession of the upper margins of a glacier is the key symptoms to observe (Kääb *et al.*, 2002; Pelto, 2010; Carturan *et al.*, 2013). Glaciers disappearing and fragmenting have become a common reporting category of glacier inventories, both indicating disequilibrium with climate (Tennant *et al.*, 2012; Kulkarni and Karyakarte, 2014; Carturan *et al.*, 2013).

1.3.3 Accumulation Zone Changes

Paul *et al.* (2004) utilized satellite imagery to identify nonuniform changes in glacier geometry, emerging rock outcrops, disintegration, and tributary separation to determine collapse versus a dynamic (equilibrium) response to climate change. It has become practical to examine the terminus and areal extent change of all glaciers in the region (Bajracharya and Mool, 2009; Paul *et al.*, 2004). This does quantify the extent of the retreat, but not the nature of the equilibrium or disequilibrium response. Identifying disequilibrium requires identification of significant thinning in the accumulation zone, which will occur if a persistent accumulation zone is lacking. Accumulation zone thinning is evidenced by the emergence of rock outcrops in the accumulation zone, changes in the accumulation zone margin, and reduced crevassing (Paul *et al.*, 2004; Pelto, 2010).

1.3.4 Terminus Response Factors

Glacier terminus response to climate depends on several dynamic features primarily: the existence of debris cover, having a calving terminus, and being a surging glacier. These dynamic features do not render a glacier insensitive to climate change in the long run, but can mitigate or accentuate the response in the short run.

Debris cover is common in three of the regions examined in this volume: Patagonia, New Zealand, and Himalayas. Once the debris cover in the ablation zone exceeds a thickness of 1 cm, it insulates the glacier below from the atmosphere, reducing ablation rates (Brock *et al.*, 2010). This leads to a slower thinning and less negative mass balances (Scherler, Bookhagen, and Strecker, 2011). In the long run, debris-covered glaciers exhibit the same response to climate – just delayed and subdued.

Calving glaciers lose mass by the calving of icebergs into a lake or ocean. A glacier that terminates in water leads to enhanced melting of the glacier front, which typically leads to greater calving. Calving increases with the ratio of ice thickness to water depth. The greater this ratio, the less the buoyancy of the calving front and the less the calving rate. A greater ratio would also lead to less ablation from contact with the water. The key parameter identified by Benn, Warren, and Mottram (2007) is the variation of longitudinal strain rate, which determines the crevasse depth near the calving front. Calving typically enhances retreat as it is another mechanism to lose volume. If there is a change in water depth at the calving front, and calving is reduced or increased, the glacier sensitivity to climate is reduced (Benn, Warren, and Mottram, 2007).

Surging glaciers experience periodic periods of slow and rapid flows. The periods of rapid flow are much shorter in duration than the slow-flow periods. During the periods of slow flow, a glacier thickens, which promotes higher velocity and also leads to hydrologic changes at the glacier bed that is crucial to generating a period of rapid flow. During the slow phase, a glacier's response to climate change is often enhanced, while during the rapid-flow periods a surging glacier is less sensitive to climate conditions.

1.4   Glacier Runoff

Glaciers act as natural reservoirs storing water in a frozen state instead of behind a dam. Alpine Watersheds are comprised of rainfall-dominated (pluvial), snowmelt-dominated (nival), and glacier melt-dominated segments. Glacially fed streams peak during late summer, July and August in the Northern Hemisphere, and January and February in the Southern Hemisphere, during peak glacier melt (Fig. 1.2) (Fountain and Tangborn, 1985; Dery *et al.*, 2009). The loss of glaciers from a watershed would result in reduced streamflow primarily during late summer minimum flow periods (Pelto, 2008; Nolin *et al.*, 2010; Stahl and Moore, 2006).

1.5   Climate Change and Impact of Runoff

Glacier runoff is the product of glacier area and glacier melt rate. If the percent decline in glacier area is greater than the percent increase in melt rate, runoff declines. If melt rate percentage increase is greater than the percentage of area loss, glacier runoff will increase. The point at which glacier runoff begins to decline due to area loss is considered the peak glacier runoff. In some areas, peak runoff has already occurred. Bliss, Hock, and Radić (2014) observed that of 18 glaciated alpine areas, most regions exhibit a fairly steady decline in runoff, demonstrating that they have passed their peak runoff. Moore *et al.* (2009) found significant declines in stream discharge from glaciers in British Columbia during late summer. Pelto (2011) found the Skykomish River glacier runoff had peaked by 2006. In the Swiss Alps, peak runoff overall has not occurred (Huss, 2011). Isaak *et al.* (2011) have found significant and ubiquitous warming of streams in the Pacific Northwest during the summer in unregulated rivers from 1980 to 2009 of 0.12 °C/decade. The combination of reduced flow and temperature rise will be an ongoing threat to aquatic life (Grah and Beaulieu, 2013).

The impact from the Nooksack River, WA, serves as an example (Pelto, 2015a). Discharge and water temperature are measured in each river. The South Fork has 0% glacier cover, Middle Fork 2.1%

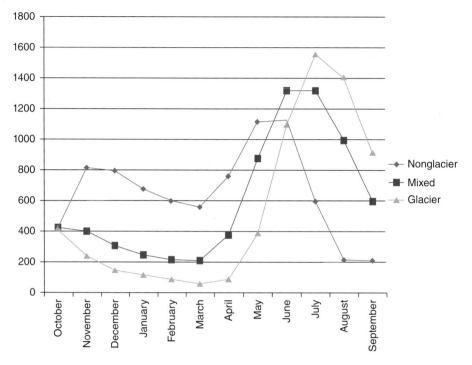

Figure 1.2 Comparison of hydrographs for nonglacier- and glacier-fed watersheds.

glacier cover, and North Fork 6.1% glacier cover, allowing differentiation of glacier impact. In addition, streamflow is measured directly below one glacier in the North Fork, Sholes Glacier. To distinguish the different discharge and thermal responses, the comparison was made during the most stressful period – late summer warm weather events. For the 2009–2013 period, 12 warm weather events were identified. The mean increase in air temperature during the warm weather events from prior to their beginning was 7 °C. Warm weather events consistently generate a significant increase in stream water temperature only in the nonglaciated South Fork Basin; the mean increase was 3.2 °C (Table 1.1). Increased glacier discharge largely offset the impact of increased air temperature on stream water temperature during the warm weather events, leading to a mean change of 1.1 °C in the North Fork and 1.0 °C in the Middle Fork.

Table 1.1 Mean response of Nooksack River watershed to the 14 warm weather events from 2009 to 2013.

Basin	Air temperature change (°C)	Stream temperature change (°C)	Stream discharge change (%)
South Fork Nooksack	+8	+3.4	−15
North Fork Nooksack	+8	+1.1	+23
Middle Fork Nooksack	+8	+1.0	+16

The change in air temperature is assessed as the rise in the mean daily temperature at the Middle Fork Nooksack SNOTEL site.

For discharge during the same warm weather events, a 15% increase is set as the key threshold for a significant response to each warm weather event. This threshold was chosen as only significant rain or melt events generate this large a change in daily flow. For the North Fork, 11 of 12 warm weather events exceeded this limit, in the Middle Fork 8 of 12 events had a significant response, and for the South Fork none of the 12 events led to a 15% flow increase. The average discharge change for the warm weather events are +26% in the North Fork, +19 % in the Middle Fork, and −16% in the South Fork (Table 4.3). It is apparent that warm weather events increase glacier melt enhancing flow in the North Fork, and in a basin without glacier runoff, South Fork, the hydrologic system consistently experiences reduced discharge.

References

Arendt, A.A. *et al.* (2012) *Randolph Glacier Inventory [v2.0]: A Dataset of Global Glacier Outlines.* Global Land Ice Measurements from Space, Boulder, CO, http://www.glims.org/RGI/randolph.html (last accessed 24 May 2016).

Bajracharya, S.R. and Mool, P. (2009) Glaciers, glacial lakes and glacial lake outburst floods in the Mount Everest region, Nepal. *Annals of Glaciology*, **50** (53), 81–86. doi: 10.3189/172756410790595895

Benn, D.I., Warren, C.R., and Mottram, R.H. (2007) Calving processes and the dynamics of calving glaciers. *Earth Science Reviews*, **82** (3–4), 143–179. doi: 10.1016/j.earscirev.2007.02.002

Bliss, A., Hock, R., and Radić, V. (2014) Global response of glacier runoff to twenty-first century climate change. *Journal of Geophysical Research, Earth Surface*, **119** (4), 717–730. doi: 10.1002/2013JF002931

Brock, B.W., Mihalcea, C., Kirkbride, M. *et al.* (2010) Meteorology and surface energy fluxes in the 2005–2007 ablation seasons at the Miage debris-covered glacier, Mont Blanc Massif, Italian Alps. *Journal of Geophysical Research*, **115**, D09106. doi: 10.1029/2009JD013224

Carturan, L., Filippi, R., Seppi, R. *et al.* (2013) Area and volume loss of the glaciers in the Ortles-Cevedale group (Eastern Italian Alps): controls and imbalance of the remaining glaciers. *The Cryosphere*, 7, 267–319.

Dery, S., Stahl, K., Moore, R. *et al.* (2009) Detection of runoff timing changes in pluvial, nival and glacial rivers of western Canada. *Water Resources Research*, **45**. doi: 10.1029/2008WR006975

Fischer, M., Huss, M., Barboux, C., and Hoelzle, M. (2014) The new Swiss Glacier Inventory SGI2010: Relevance of using high resolution source data in areas dominated by very small glaciers. *Arctic, Antarctic, and Alpine Research*, **46**, 935–947. doi: 10.1657/1938-4246-46.4.933

Forel, F. (1895) Les variations periodiques des glaciers. Discours preliminaire. *Archives des Sciences Physiques et Naturelles*, **34**, 209–229.

Fountain, A. and Tangborn, W. (1985) The Effect of Glaciers on Streamflow Variations. *Water Resources Research*, **21**, 579–586.

GISTEMP Team (2015) *GISS Surface Temperature Analysis (GISTEMP).* NASA Goddard Institute for Space Studies, http://data.giss.nasa.gov/gistemp/ (dataset accessed 25 October 2015).

Grah, O. and Beaulieu, J. (2013) The effect of climate change on glacier ablation and baseflow support in the Nooksack River basin and implications on Pacific salmonid species protection and recovery. *Climatic Change.* doi: 10.1007/s10584-013-0747-y

Haeberli, W., Cihlar, J., and Barry, R. (2000) Glacier monitoring within the Global Climate Observing System. *Annals of Glaciology*, **31**, 241–246.

Huss, M. (2011) Present and future contribution of glacier storage change to runoff from macroscale drainage basins in Europe. *Water Resources Research*, **47**, W07511. doi: 10.1029/2010WR010299

Immerzeel, W.W., Beek van, L.P., and Bierkens, M.F. (2010) Climate change will affect the Asian water towers. *Science*, **328**, 1382–1385.

Isaak, D.J., Wollrab, S., Horan, D., and Chandler, G. (2011) Climate change effects on stream and river temperatures across the northwest U.S. from 1980–2009 and implications for salmonid fishes. *Climate Change*. doi: 10.1007/s10584-011-0326-z

Johannesson, T., Raymond, C., and Waddington, E. (1989) Time-scale for adjustment of glacier to changes in mass balance. *Journal of Glaciology*, **35** (121), 355–369.

Kääb, A., Paul, F., Maisch, M. *et al.* (2002) The new remote-sensing-derived Swiss Glacier Inventory: II. First results. *Annals of Glaciology*, **34**, 362–366.

Kulkarni, A. and Karyakarte, Y. (2014) Observed changes in Himalayan glaciers. *Current Science*, **106** (2), 237–244.

Marzeion, B., Jarosch, A., and Hofer, M. (2012) Past and future sea-level change from the surface mass balance of glaciers. *The Cryosphere*, **6**, 1295–1322. doi: 10.5194/tc-6-1295-2012

Mernild, S., Pelto, M., Malmros, J. *et al.* (2013) Identification of snow ablation rate, ELA, AAR and net mass balance using transient snowline variations on two Arctic glaciers. *Journal of Glaciology*, **59**, 649–659. doi: 10.3189/2013JoG12J221

Moore, R.D., Fleming, S.W., Menounos, B. *et al.* (2009) Glacier change in western North America: influences on hydrology, geomorphic hazards and water quality. *Hydrological Processes*, **23**, 42–61.

Neckel, N., Kropacek, J., Bolch, T., and Hochschild, V. (2014) Glacier mass changes on the Tibetan Plateau 2003–2009 derived from ICESat laser altimetry measurements. *Environmental Research Letters*, **9** (1), 014009. doi: 10.1088/1748-9326/9/1/014009

Nolin, A.W., Phillippe, J., Jefferson, A., and Lewis, S.L. (2010) Present-day and future contributions of glacier runoff to summertime flows in a Pacific Northwest watershed: implications for water resources. *Water Resources Research*, **46** (12), W12509.

Oerlemans, J. (2005) Extracting a climate signal from 169 glacier records. *Science*, **308** (5722), 675–677. doi: 10.1126/science.1107046

Østrem, G. (1975) ERTS data in glaciology—an effort to monitor glacier mass balance from satellite imagery. *Journal of Glaciology*, **16**, 403–415.

Paterson, W.S.B. (1994) *The Physics of Glaciers*, 3rd edn, Pergamon, Oxford, UK, p. 480.

Paul, F., Kääb, A., Maisch, M. *et al.* (2004) Rapid disintegration of Alpine glaciers observed with satellite data. *Geophysical Research Letters*, **31**, L21402.

Pelto, M. (2008) Glacier annual balance measurement, forecasting and climate correlations, North Cascades, Washington 1984–2006. *The Cryosphere*, **2**, 13–21.

Pelto, M. (2010) Forecasting temperate Alpine glacier survival from accumulation zone observations. *The Cryosphere*, **3**, 323–350.

Pelto, M.S. (2011) Skykomish river, Washington: impact of ongoing glacier retreat on streamflow. *Hydrological Processes*, **25** (21), 3267–3371.

Pelto, M. (2015a) Climate driven retreat of mount Baker glaciers and changing water resources, in *SpringerBriefs in Climate Studies*, Springer International Publishing. doi: 10.1007/978-3-319-22605-7

Pelto, M.S. (2015b) Alpine glaciers (in "State of the Climate in 2014"). *Bulletin of the American Meteorological Society*, **96** (7), S19–S20.

Pelto, M.S. and Hedlund, C. (2001) Terminus behavior and response time of North Cascade glaciers, Washington, U.S.A. *Journal of Glaciology*, **47** (158), 497–506. doi: 10.3189/172756501781832098

Pfeffer, W.T. *et al.* (2014) The Randolph Glacier Inventory: a globally complete inventory of glaciers. *Journal of Glaciology*, **60** (221), 537–552. doi: 10.3189/2014JoG13J176

Radić, V., Bliss, A., Beedlow, A.C. *et al.* (2013) Regional and global projections of twenty-first century glacier mass changes in response to climate scenarios from global climate models. *Climate Dynamics*, **42** (1–2), 37–58. doi: 10.1007/s00382-013-1719-7

Radić, V. and Hock, R. (2014) Regional and global volumes of glaciers derived from statistical upscaling of glacier inventory data. *Journal of Geophysical Research*, **115** (F1), F01010. doi: 10.1029/2009JF001373

Schaner, N., Voisin, N., Nijssen, B., and Lettenmaier, D. (2012) The contribution of glacier melt to streamflow. *Environmental Research Letters*, **7**. doi: 10.1088/1748-9326/7/3/034029

Scherler, D., Bookhagen, B., and Strecker, M. (2011) Spatially variable response of Himalayan glaciers to climate change affected by debris cover. *Nature Geoscience*, **4** (3), 156–159. doi: 10.1038/ngeo1068

Schwitter, M.P. and Raymond, C. (1993) Changes in the longitudinal profile of glaciers during advance and retreat. *Journal of Glaciology*, **39** (133), 582–590.

Sharp, M., Wolken, G., Burgess, D. *et al.* (2015) Glaciers and ice caps outside Greenland, in "State of the Climate in 2014". *Bulletin of the American Meteorological Society*, **96** (7), S135–S137.

Stahl, K. and Moore, D. (2006) Influence of watershed glacier coverage on summer streamflow in British Columbia, Canada. *Water Resources Research*, **42** (W6), W06201. doi: 10.1029/2006WR005022

Tennant, C., Menounos, B., Wheate, R., and Clague, J.J. (2012) Area change of glaciers in the Canadian Rocky Mountains, 1919 to 2006. *The Cryosphere*, **6** (6), 1541–1552. doi: 10.5194/tc-6-1541-2012

WGMS (2011) *Glacier Mass Balance Bulletin No. 11 (2008–2009)*. Zemp, M., Nussbaumer, S. U., GärtnerRoer, I., Hoelzle, M., Paul, F., and Haeberli, W. (eds.), ICSU(WDS)/IUGG(IACS)/UNEP/UNESCO/WMO, World Glacier Monitoring

WGMS (2012) *Fluctuations of Glaciers 2005–2010 (Vol. X)*. ICSU(WDS)/IUGG(IACS)/UNEP/UNESCO/WMO, World Glacier Monitoring Service, Zurich. doi: 10.5904/wgms-fog- 2012-11

WGMS (2013) *Glacier Mass Balance Bulletin No. 12 (2010–2011)*. ICSU(WDS)/IUGG(IACS)/UNEP/UNESCO/WMO, World Glacier Monitoring Service, Zurich, doi: 10.5904/wgms-fog-2013-11

Zemp, M., Hoelzle, M., and Haeberli, W. (2009) Six decades of glacier mass-balance observations: a review of the worldwide monitoring network. *Annals of Glaciology*, **50** (50), 101–111. doi: 10.3189/172756409787769591

Zemp, M. *et al.* (2015) Historically unprecedented global glacier decline in the early 21st century. *Journal of Glaciology*, **61** (228), 745–763. doi: 10.3189/2015JoG15J017

2

Glacier Mass Balance

Overview

Crucial to the existence and survival of a glacier is its surface mass balance – the difference between accumulation and ablation (sublimation and melting). Climate change may cause variations in both temperature and snowfall, causing changes in the surface mass balance. In turn, the cumulative changes in mass balance control a glacier's dynamic and terminus behavior and are the most sensitive climate indicators on a glacier. From 1980 to 2014, the mean cumulative mass loss of glaciers reporting mass balance to the World Glacier Monitoring Service is −16.8 m – a mean annual loss of −0.48 m. This includes 31 consecutive years of negative mass balances.

Mass balance is reported in meters of water equivalent (w.e.). This represents the average thickness gained (positive balance) or lost (negative balance) from the glacier during that particular year. Annual mass balance is measured by determining the amount of snow accumulated during winter, and later measuring the amount of snow and ice removed by melting in the summer. The difference between these two parameters is the mass balance. If the amount of snow accumulated during the winter is larger than the amount of melted snow and ice during the summer, the mass balance is positive and the glacier has increased in volume. On the other hand, if the melting of snow and ice during the summer is larger than the supply of snow in the winter, the mass balance is negative and glacier volume decreases.

The recent glacier retreat is a reflection of strongly negative mass balances over the last 30 years (Zemp, Hoelzle, and Haeberli, 2009). The cumulative mass balance loss since 1980 is 16.8 m w.e., the equivalent of cutting an 18.5-m-thick slice off the top of the average glacier (Fig. 2.1). The trend is remarkably consistent from region to region (WGMS, 2015). WGMS mass balance based on 37 reference glaciers with a minimum of 30 years of record is not appreciably different at 16.4 m w.e. The decadal mean annual mass balance was −221 mm in the 1980s, 389 mm in the 1990s, and −726 mm in the 2000s. The declining mass balance trend during a period of retreat indicates that alpine glaciers are not approaching equilibrium and retreat will continue to be the dominant terminus response. The recent rapid retreat and prolonged negative balances have led to some glaciers disappearing and others fragmenting (Pelto, 2010; Carturan et al., 2013).

To determine mass balance in the accumulation zone, snowpack depth is measured using probing, snowpits, ground-penetrating radar, or crevasse stratigraphy. Crevasse stratigraphy makes use of annual layers revealed on the wall of a crevasse (Fig. 2.2). Akin to tree rings, these layers are formed due to summer dust deposition and other seasonal effects (Post and LaChapelle, 1971; Pelto and Riedel, 2001; Purdie et al., 2011). The advantage of crevasse stratigraphy is that it provides a two-dimensional measurement of the snowpack layer – not a point measurement. It is also usable in depths where probing or snowpits are not feasible. In temperate glaciers, the insertion resistance of

Recent Climate Change Impacts on Mountain Glaciers, First Edition. Mauri Pelto.
© 2017 John Wiley & Sons, Ltd. Published 2017 by John Wiley & Sons, Ltd.

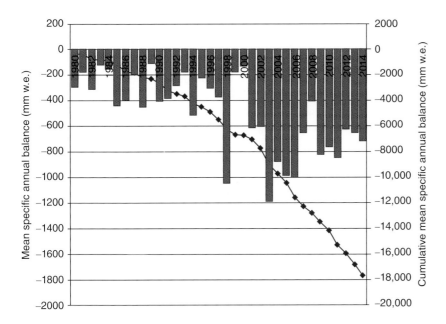

Figure 2.1 Annual and cumulative balance of the WGMS reference glaciers.

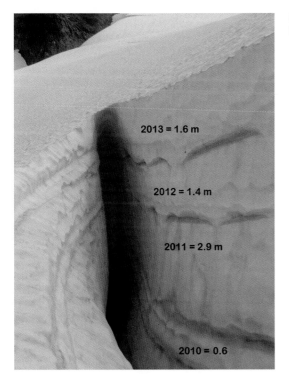

Figure 2.2 Annual layers in a crevasse on Lynch Glacier, North Cascade Range, Washington.

a probe increases abruptly when its tip reaches ice that was formed the previous year (Østrem and Brugman, 1991). The probe depth is a measure of the net accumulation above that layer. Snowpits dug through the past winter's residual snowpack are used to determine the snowpack depth and density. The snowpack's mass balance is the product of density and depth (Østrem and Brugman, 1991). Ground-penetrating radar is a new technique that can provide a much greater spatial coverage of snow depth measurement (McGrath *et al.*, 2015). Regardless of depth measurement technique, the observed depth is multiplied by the snowpack density to determine the accumulation in water equivalent. It is necessary to measure the density in the spring as snowpack density varies. In some regions, snowpack density is uniform by the end of the melt season, and probing has even greater utility. Measurements of snowpack density completed at the end of the ablation season yield consistent values for a particular area on temperate alpine glaciers and need not be measured every year. In other regions, meltwater is refrozen as superimposed ice or ice lenses within the firn and snowpack and density must be measured.

In the ablation zone, ablation measurements are made using stakes inserted vertically into the glacier either at the end of the previous melt season or the beginning of the current one. The length of the stake exposed by melting ice is measured at the end of the melt (ablation) season. Most stakes must be replaced each year or even midway through the summer (Østrem and Brugman, 1991). Ablation near the snowline can also be assessed using the migration of the transient snowline with time if the balance gradient is known (Mernild *et al.*, 2013). Recently, ablation is also being monitored using repeat Light Detection and Ranging (LIDAR) observations (Helfricht *et al.*, 2014).

Geodetic mass balance methods are employed for the determination of mass balance of glacier based on mapping of glacier area and surface elevation. Maps of a glacier made at two different points in time can be compared and the difference in glacier thickness observed can be used to determine the mass balance over a span of years. This is typically done by generation of digital elevation model (DEM) for a glacier from different mapping campaigns. Sometimes, the earliest data for the glacier surface profiles is from aerial images that were used to make topographical maps, and more recent mapping is supplemented with remote sensing, radar altimetry, and laser altimetry. Aerial mapping or photogrammetry is now used to cover larger glaciers and ice caps such as those found in Antarctica and Greenland; however, because of the problems in establishing accurate ground control points in mountainous terrain, and correlating features in snow and where shading is common, elevation errors are typically not less than 10 m. LIDAR provides a measurement of the elevation of a glacier along a specific path, for example, the glacier centerline. The difference of two such measurements is the

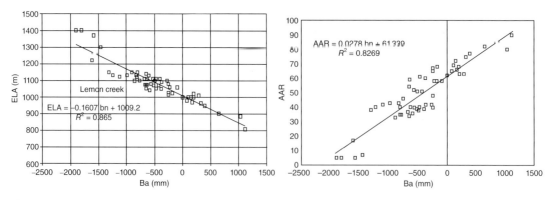

Figure 2.3 Example of the relationship between annual balance (Ba), accumulation area ratio (AAR), and equilibrium line altitude (ELA) as reported to the WGMS.

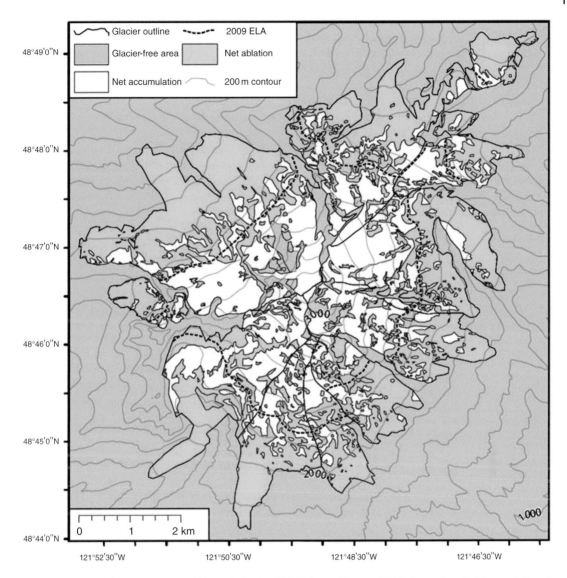

Figure 2.4 Accumulation area map of Mount Baker in 2009. (Pelto and Brown (2012). Reproduced with permission of Springer.)

change in thickness, which provides mass balance over the time interval between the measurements (Rees and Arnold, 2007). The value of geodetic programs has been providing an independent check of traditional mass balance work, by comparing the cumulative changes over 10 or more years. As airborne and terrestrial LIDAR methodology is further refined, its use in annual mass balance work is becoming vital. Zemp, Hoelzle, and Haeberli (2009) emphasized the need for complimentary geodetic and field-based mass balance measurement. This has led to numerous reassessments of long-term mass balance records (Zemp *et al.*, 2015).

The importance of monitoring glacier mass balance was recognized during the International Geophysical Year (IGY) in 1957. For the IGY, a number of benchmark glaciers around the world were chosen where mass balance would be monitored. This network has proven valuable, but in many areas the number of glacier is limited; for example, there is just one benchmark glacier in the North Cascades and in the conterminous United States – South Cascade Glacier (Fountain *et al.,* 1991). Glacier mass balance varies due to geographic characteristics such as aspect, elevation, and location with respect to prevailing winds. Since no single glacier is representative of all others to understand the causes and nature of changes in glacier surface mass balance throughout a mountain range, it is necessary to monitor a significant number of glaciers, which has been done in the North Cascades, WA (Fountain *et al.,* 1991; Pelto and Riedel, 2001).

At regional scales and on specific glaciers, there are two common proxies for assessing mass balance without detailed observations. They are the accumulation area ratio (AAR) and equilibrium line altitude (ELA) that can be derived from satellite imagery (Østrem, 1975; Racoviteanu, Williams and Barry, 2008). The ELA is the elevation at which ablation equals accumulation; on temperate alpine glaciers, this is coincident with the snowline at the end of the melt season. The ELA is not typically an easily discernible line or elevation on many smaller or steep alpine glaciers, due to variability of snow accumulation from impacts of wind and avalanche redistribution. Ascribing the ELA is also difficult when a recent DEM is not available, particularly when glacier surface elevation is changing rapidly. AAR is a more accurately determined parameter and a better proxy in this case. Rabatel *et al.* (2008) and Dyurgerov (1996) developed methods to derive mass balance from long-term AAR observations. The AAR-Ba method has proven reliable (Hock *et al.,* 2007; Racoviteanu, Williams and Barry, 2008; Pelto and Brown, 2012). The WGMS (2008, 2013) has adopted the reporting of AAR with all mass balance values and plotting the relationship for each glacier (Fig. 2.3). AAR is a parameter that can be evaluated using satellite imagery over a large region, providing an efficient mechanism for the utilization of AAR-based mass balance determination on numerous glaciers (Fig. 2.4).

Going forward, the ability to map the areal extent and elevation change of glaciers using remote sensing will provide a more spatially comprehensive data set of glacier change, sea-level contribution by glaciers, and potential glacier runoff contribution.

References

Carturan, L., Baroni, C., Becker, M. *et al.* (2013) Decay of a long-term monitored glacier: Careser Glacier (Ortles-Cevedale, European Alps). *The Cryosphere,* **7**, 1819–1838. doi: 10.5194/tc-7-1819-2013

Dyurgerov, M. (1996) Substitution of long term mass balance data by measurements of one summer. *Gletscherkd. Glazialgeol.,* **32**, 177–184.

Fountain, A., Trabant, D., Bruggman, M. *et al.* (1991) Glacier mass balance standards. *Eos,* **72** (46), 511–514.

Helfricht, K., Kuhn, M., Keuschnig, M., and Heilig, A. (2014) Lidar snow cover studies on glaciers in the Ötztal Alps (Austria): comparison with snow depths calculated from GPR measurements. *The Cryosphere,* **8**, 41–57. doi: 10.5194/tc-8-41-2014

Hock, R., Koostra, D., and Reijmeer, C. (2007) Deriving glacier mass balance from accumulation area ratio on Storglaciären, Sweden, in *Glacier Mass Balance Changes and Meltwater Discharge IAHS,* vol. **318**, pp. 163–170.

McGrath, D., Sass, L., O'Neel, S. *et al.* (2015) End-of-winter snow depth variability on glaciers in Alaska. *Journal of Geophysical Research, Earth Surface,* **120**, 1530–1550. doi: 10.1002/2015JF003539

Mernild, S., Pelto, M., Malmros, J. *et al.* (2013) Identification of snow ablation rate, ELA, AAR and net mass balance using transient snowline variations on two Arctic glaciers. *Journal of Glaciology,* **59**, 649–659. doi: 10.3189/2013JoG12J221

Østrem, G. (1975) ERTS data in glaciology—an effort to monitor glacier mass balance from satellite imagery. *Journal of Glaciology*, **16**, 403–415.

Østrem, G. and Brugman, M. (1991) Mass Balance Measurement Techniques, a Manual for Field and OfficeWork, National Hydrological Research Institute (NHRI), *Science Report, 4*, Saskatoon, Canada, 224 pp.

Pelto, M. (2010) Forecasting temperate alpine glacier survival from accumulation zone observations. *The Cryosphere*, **4**, 67–75.

Pelto, M.S. and Brown, C. (2012) Mass balance loss of Mount Baker, Washington glaciers 1990–2010. *Hydrological Processes*, **26** (17), 2601–2607.

Pelto, M.S. and Riedel, J. (2001) The spatial and temporal variation of mass balance on North Cascade glaciers. *Hydrological Processes*, **15**, 3461–3472.

Post, A. and LaChapelle, E. (1971) *Glacier Ice*, University of Washington Press, Seattle, WA.

Purdie, H., Anderson, B., Lawson, W., and Mackintosh, A. (2011) Controls on spatial variability in snow accumulation on glaciers in the Southern Alps, New Zealand; as revealed by crevasse stratigraphy. *Hydrological Processes*, **25**, 54–63. doi: 10.1002/hyp.7816

Rabatel, A., Dedieu, J., Thibert, E. *et al.* (2008) Twenty-five years of equilibrium-line altitude and mass balance reconstruction on the Glacier Blanc, French Alps (1981–2005), using remote-sensing method and meteorological data. *Journal of Glaciology*, **54**, 307–314.

Racoviteanu, A., Williams, M., and Barry, R. (2008) Optical remote sensing of glacier characteristics: a review with focus on the Himalaya. *Sensors*, **8** (5), 3355–3383. doi: 10.3390/s8053355

Rees, W. and Arnold, N. (2007) Mass balance and dynamics of a valley glacier measured by high-resolution LiDAR. *Polar Record*, **43**, 311–319. doi: 10.1017/S0032247407006419

WGMS (2008) *Fluctuations of glaciers 2000–2005 (Vol. IX)*. ICSU (FAGS)/IUGG–(IACS)/UNEP/UNESCO/WMO, World Glacier Monitoring Service, Zurich. doi: 10.5904/wgms-fog-2008-12

WGMS (2013) *Glacier Mass Balance Bulletin No. 12 (2010–2011)*. ICSU(WDS)/IUGG(IACS)/UNEP/UNESCO/WMO, World Glacier Monitoring Service, Zurich. doi: 10.5904/wgms-fog-2013-11

WGMS (2015) *Latest Glacier Mass Balance Data*. http://www.geo.uzh.ch/microsite/wgms/mbb/sum13.html (accessed May 2015).

Zemp, M., Hoelzle, M., and Haeberli, W. (2009) Six decades of glacier mass-balance observations: a review of the worldwide monitoring network. *Annals of Glaciology*, **50**, 101–111. doi: 10.3189/172756409787769591

Zemp, M. *et al.* (2015) Historically unprecedented global glacier decline in the early 21st century. *Journal of Glaciology*, **61** (228), 745–763. doi: 10.3189/2015JoG15J017

3

Juneau Icefield

Overview

Southern Alaska is one of the most glaciated alpine regions on Earth. Volume losses in the region have been a key contributor to global sea level in the last century (Berthier *et al.*, 2010). In this chapter we examine the Juneau Icefield as an example of what has been occurring in the region. Larsen *et al.*, (2007) demonstrate that losses on the Juneau Icefield are commensurate with losses across the region. The Juneau Icefield is east of Juneau, Alaska, with 15 primary outlet glaciers terminating in both Alaska and British Columbia. The icefield has a temperate maritime climate dominated by the passage of frequent cyclonic systems, with precipitation enhanced by orographic processes. The icefield covering more than 4000 km² is not as well known as the Glacier Bay region just north and west. It owes its existence to the frequent cyclonic passages from the Gulf of Alaska. The Juneau Icefield Research Program (JIRP) (2015) has been examining the glaciers of the Juneau Icefield since 1946. Until the NASA Landsat program began capturing satellite images, field measurements and aerial observations were the only means to observe the glaciers of the icefield. For more than 40 years, it was Maynard Miller, University of Idaho, who led this expedition that has trained so many of today's glaciologists. Today it is Jeff Kavanaugh, University of Alberta, who leads this expedition. Given the difficult weather conditions that produce the icefield, this was not a task that could be done comprehensively. Focus was given to the Lemon Creek and Taku Glacier. These two glaciers have the longest annual mass balance records in the United States, and the Lemon Creek Glacier is a World Glacier Monitoring Service (WGMS) reference glacier (Pelto, Kavanaugh, and McNeil, 2013). Here we examine the changes from the August 17, 1984, Landsat 5 image to the June 21, 2013, image (LC80570192013172LNG00) obtained from the newly launched Landsat 8 (Fig. 3.1). The August 17, 1984, image (LT50570191984230PAC00) is the oldest Landsat image that I consider of top quality. I was on the Llewellyn Glacier with JIRP on the east side of the icefield the day this image was taken. Reference is also made to the changes since the 1948 aerial photographs used to complete the USGS topographic map. Landsat 5 was launched in 1984, and Landsat 8 in 2013. The Landsat images have become a key resource in the examination of the mass balance of these glaciers (Pelto, 2011).

Figure 3.1 shows the reference image of the entire icefield that indicates the location of the 15 main glaciers we focus on here. The chart shown in Fig. 3.2 indicates the amount of terminus change from 1984 to 2013: 14 glaciers have retreated and 1 has advanced.

Figure 3.3 displays a detailed glacier comparison of the terminus, with the images from 1984 (always on the left) and 2013 (on the right): the 1984 margin is marked with red dots and the 2013 with yellow dots. The satellite images were first overlain in GIS, and the terminus change was based on

Recent Climate Change Impacts on Mountain Glaciers, First Edition. Mauri Pelto.
© 2017 John Wiley & Sons, Ltd. Published 2017 by John Wiley & Sons, Ltd.

Figure 3.1 June 21, 2013, Landsat 8 image indicating outlet glaciers of the Juneau Icefield, Alaska. N = Norris, L = Lemon Creek, M = Mendenhall, H = Herbert. E = Eagle, G = Gilkey, A = Antler, F = Field, Ll = Llewellyn, Tu = Tulsequah, Tw = Twin, and T = Taku. (Landsat, US Geological Survey.)

three measurements – one at the glacier terminus midpoint and one each halfway to the margin from the midpoint. An exception is the Taku Glacier, which is based on the JIRP field measurement mean, and the Llewellyn Glacier, where three measurements are made on each of the two termini; the average is then rounded to the nearest 100 m. The ongoing retreat reflects the long-term negative mass balance of the glaciers with the exception of the Taku Glacier. Taku Glacier was a tidewater glacier until 1948 and is still adjusting to being a land-terminating glacier. The ongoing warming of our globe will continue to lead to retreat. The glaciers are all fed from the central portion of the icefield that always has a large snow-covered area even at the end of recent warm summers.

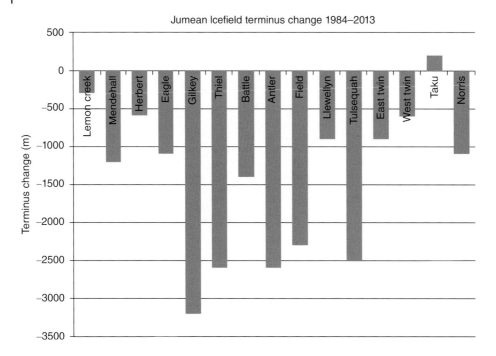

Figure 3.2 The 1984–2013 chart of terminus change of individual glaciers from 1984 to 2013 (see individual images provided later for the observed changes).

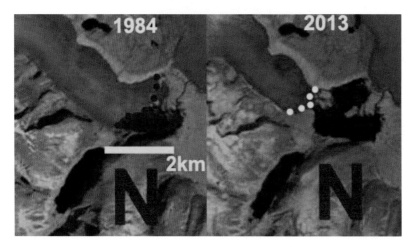

Figure 3.3 Norris Glacier terminus change from 1984 to 2013. Red and yellow dots indicate the 1984 and the 2013 termini, respectively.

The 2015 season will be of interest, since the area had a remarkably warm yet wet winter. This will lead to high ablation at lower elevations, likely a higher snowline than usual, but above the Vaughan Lewis Icefall, will those warm wet events dump snow? The 2014 winter season was warm and the snowline seen in the 2014 satellite imagery was at 1500 m, yet snowpack at 1760 m on the Vaughan Lewis Glacier was 3.3 m deep in late July. This has been the case in the past with warm wet winters

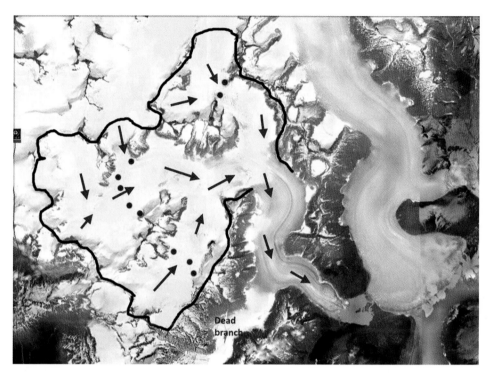

Figure 3.4 Norris Glacier overview, illustrating flow paths, snowline, and glacier boundaries.

featuring heavy snow above 1600 m on the icefield. JIRP will be in the field looking to help answer this question in 2015.

3.1 Norris Glacier

I have had the chance to cross this glacier on ski a half dozen times as part of JIRP, and it was seldom an easy traverse, hence the names death valley and dead branch for portions of the glacier. The view of the entire glacier in Google Earth illustrates the direction of flow and accumulation sources (blue arrows), typical snowline (red dots), and glacier boundary (black line) (Fig. 3.4).

Norris Glacier began retreating before 1890 and has continuously retreated 2050 m from its 19th to 20th maximum achieved around 1915. The glacier ended in a lake referred to here as Norris Lake from 1948 until 2007. In 1948 the USGS map indicated that Norris Lake was a narrow 200–500 m wide lake and the glacier terminus was only 200–700 m from Grizzly Bar, increasing with distance south along margin an outwash plain built by the glacier beyond its advanced terminus position.

From 1948 to 1975 the glacier retreated 250 m, and Norris Lake had expanded but still had considerable ice along its western margin. The terminus was still extending a kilometer downvalley of the outlet of Glory Lake (Fig. 3.3).

In 1998 I had the opportunity to fly over the glacier at the end of the field season. A series of images indicates the trimline that had developed with the recent glacier thinning for the last 50 years. Norris Lake had expanded to a length of 1.2 km on the north shore and 1.9 km on the south shore from

Figure 3.5 Trimline above Norris Glacier in 1998 and the terminus in Norris Lake.

Grizzly Bar. There are numerous icebergs in Norris Lake in 1998, and the glacier reached the lake across most of its western half (Fig. 3.5).

In 2010 the glacier had retreated from this lake. From 1984 to 2013 Norris Glacier has retreated 1100 m. The glacier terminus that has been ending in a proglacial lake for the last 40 years is now almost entirely grounded. Since 1984 the northern half of this lake has formed and the long-term lake development is ending.

An examination of the icefall feeding the glacier terminus area indicates considerable melt out of crevasse features, indicating that flow is not that vigorous through the icefall above the terminus at 300 m elevation, suggesting retreat will continue. The trimline has increased from 20 to 50 m above the ice surface since 1975.

3.2 Lemon Creek Glacier

Lemon Creek Glacier has a long-term mass balance record that indicates more than 15 m of thinning from 1984 to 2014 (Fig. 3.6). This thinning is more dramatic than the 300 m retreat that has occurred. The yellow arrow indicates a tributary that no longer connects to the glacier. The JIRP mass balance program from 1953 to 2013 provides a continuous 61-year record (Heusser and Marcus, 1964; Miller and Pelto, 1999; Pelto, Kavanaugh, and McNeil, 2013). This is one of the nine American glaciers selected in a global monitoring network during the IGY, 1957–1958, and one of only two where measurements have continued.

The long-term record is a cumulative ice loss of −13.9 m (12.7 m w.e.) from 1957 to 1989, −19.0 m (−17.1 m w.e.) from 1957 to 1995, and −24.4 m (−22.0 m w.e.) from 1957 to 1998. The mean annual balance of the 61-year record is $-0.43 \, \text{m a}^{-1}$ and a loss of at least 30 m of ice thickness for the period

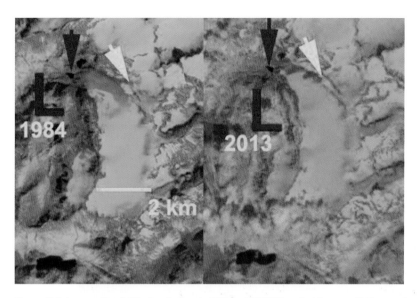

Figure 3.6 Lemon Creek Glacier change in 1984 and 2013 Landsat images. Yellow and red dots indicate the 2013 and the 1984 terminus positions, respectively. (Landsat, US Geological Survey.)

from 1953 to 2013. In the second graph the similarity with other North American glaciers is evident (Pelto, Kavanaugh, and McNeil, 2013) (Fig. 3.7).

The continued negative mass balance throughout the period has led to continued retreat since 1948. The annual balance trend indicates that despite a higher mean elevation and a higher elevation terminus, from thinning and retreat, mean annual balance has been strongly negative since 1977

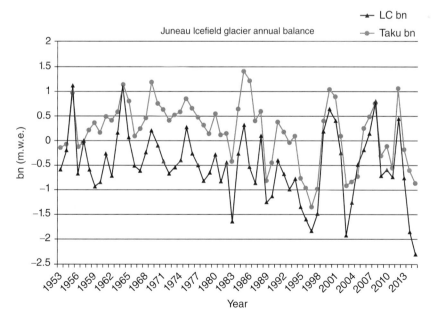

Figure 3.7 Annual mass balance record of Lemon Creek Glacier.

Figure 3.8 Lemon Creek Glacier in 2014 (note the height of the snowline). (Chris McNeil.)

$(-0.60\,\mathrm{m\,a^{-1}})$. Dramatically negative mass balances have occurred since the 1990s, with 1996, 1997, and 2003 being the only years with no retained accumulation since field observations began in 1948. This has led to an increase in the rate of thinning and terminus retreat. The glacier has experienced a terminal retreat of 1000 m during the 1953–2013 period, with the formation of a new lake before 1981 that the glacier has now retreated from. With significant negative balances in 2013 and 2014 and a record negative balance in 2015, this retreat is ongoing.

The most significant change has been that the snowline has risen above the entire glacier in several recent years, indicating the glacier cannot survive (Fig. 3.8). The continued area loss of the glacier will lead to declines in summer glacier runoff for the Lemon Creek watershed.

3.3 Mendenhall Glacier

Mendenhall Glacier is the most visited and photographed terminus in the region. The glacier in 1984 ended at the tip of a prominent peninsula in Mendenhall Lake (Fig. 3.9). In 2013 the terminus has retreated 1200 m, with an equal expansion of the lake. The lake did not exist until after 1910, and in 1984, the lake was 2700 m across indicating that, as Boyce, Motyka, and Truffer (2007) observed, the glacier had two periods of rapid retreat: one in the 1940s and the other in the 1990s. The former led to a 700 m retreat from 1948 to 1984. The latter, due to the enhanced rate of retreat since 1984, is a result of greater summer melting and a higher snowline by the end of the summer, which has averaged since 1100 m since 2003, and a greater flotation of the thinner ice (Pelto, Kavanaugh, and McNeil, 2013; Boyce, Motyka, and Truffer, 2007). In 2005 the base of the glacier was below the lake level for at least 500 m upglacier of the terminus (Boyce, Motyka, and Truffer, 2007). This suggests

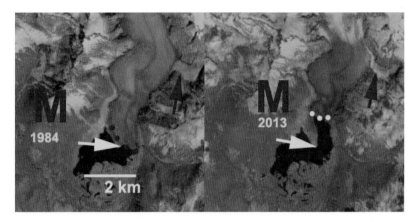

Figure 3.9 Mendenhall Glacier change in 1984 and 2013 Landsat images. Yellow and red dots indicate the 2013 and the 1984 terminus positions, respectively. (Landsat, US Geological Survey.)

that the glacier is nearing the end of the calving-enhanced retreat. It is likely that another lake basin would develop 1 km above the current terminus, where the glacier slope is quite modest. Since 2003 the equilibrium line altitude (ELA) has averaged 1225 m, which is 150 m higher than the ELA.

The red arrows indicate a tributary that decreased dramatically in width and contribution to the main glacier. This is the location of Suicide Basin, where a lake has formed during the last two summers and then rapidly drained. The USGS has installed a gage to help forecast when the lake will rapidly drain. With continued retreat, this glacier-dammed lake will cease to exist.

3.4 Herbert Glacier

Herbert Glacier is in the next valley north of Mendenhall Glacier. The glacier descended out of the mountains ending on the coastal plain in 1948. Retreat since has led to the formation of a new lake and the retreat of the glacier into the mountain valley. In 1984 we examined the terminus of this glacier, which was in the small proglacial lake at 150 m (Fig. 3.10). Herbert Glacier has retreated

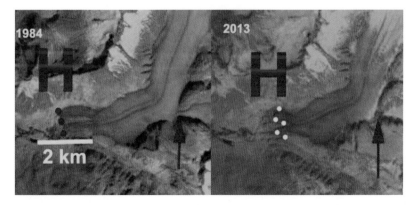

Figure 3.10 Herbert Glacier change in 1984 and 2013 Landsat images. Yellow and red dots indicate the 2013 and the 1984 terminus positions, respectively. (Landsat, US Geological Survey.)

600 m since 1984. The width of the terminus has also declined. The retreat has not been enhanced by iceberg calving as is the case at Mendenhall Glacier. The red arrow indicates a tributary that no longer feeds the main glacier. The transient snowline on the glacier has averaged 1100 m from 2003 to 2014. This leaves the glacier with an accumulation area ratio (AAR) of 0.50; too low to sustain equilibrium, retreat will continue.

3.5 Eagle Glacier

Eagle Glacier is in the next valley north of Herbert Glacier. Eagle Glacier has experienced a significant and sustained retreat since 1948. The first map image shown in Fig. 3.12 is of the glacier in 1948. At this time the glacier ended at the south end of a yet-to-be-formed glacier lake. In 1982 when I first saw the glacier and when it was mapped again by the USGS, the glacier had retreated to the north end of this 1 km long lake. In 1984 the glacier terminated at the northern edge of a still expanding proglacial lake (Fig. 3.11).

The glacier has now retreated upvalley creating a lengthening river valley. The retreat of 1100 m from 1984 to 2013 is rivaled by the width reduction of the glacier in the lower 3 km. In the USGS topographic maps based on 1948 photographs and the 2005 Google Earth image, the red line indicates the 1948 terminus, the magenta line the 1982 terminus, the green line 2005 terminus, and the orange line the 2011 terminus (Fig. 3.12).

From the 1984 to the 2005 image used in Google Earth, the glacier retreated 500 m, 21 m a^{-1}. In 2013 another 600 m of retreat had occurred. The rate of retreat has increased with time. Going back to the 1948 map, the terminus in 2011 is located where the ice was 125–175 m thick in 1948.

The more rapid retreat follows the pattern of more negative balances experienced by the glaciers of the Juneau Icefield (Pelto, Kavanaugh, and McNeil, 2013). The ELA which marks the boundary between the accumulation and the ablation zone each year is a good marker of this. As regards Eagle Glacier, to have equilibrium the glacier needs to have an ELA of 1050 m. At this elevation more than

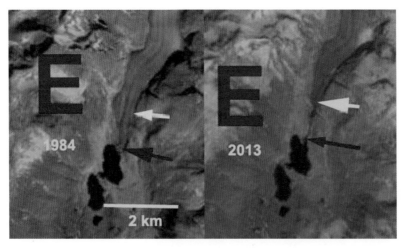

Figure 3.11 Eagle Glacier change in 1984 and 2013 Landsat images. Yellow and red arrows indicate the 2013 and the 1984 terminus positions, respectively. (Landsat, US Geological Survey.)

Figure 3.12 Google Earth image from 2011. The red line indicates the 1948 terminus, the magenta line the 1982 terminus, the green line 2005 terminus, and the orange line the 2011 terminus. (Google Earth.)

60% of the glacier is in the accumulation zone. Satellite imagery allows identification of the ELA each year. It has averaged 1175 m, which will lead to continued retreat.

3.6 Gilkey Glacier

Gilkey Glacier had begun to retreat into a proglacial lake in 1984, and the lake was still just 1 km long. A short distance above the terminus, the Gilkey was joined by the sizable tributaries of the Thiel and Battle Glaciers (Fig. 3.13). Today the map of the area is much different (Fig. 3.14).

Arrow #1 indicates the Gilkey Glacier terminus area (Fig. 3.15). In 2014 the main glacier terminus has retreated 3200 m, and the lake is now 4 km long – a lake that did not exist in the USGS maps from 1948. Thiel and Battle Glaciers have separated from the Gilkey Glacier and from each other. Thiel Glacier retreated 2600 m from its junction with Gilkey Glacier from 1984 to 2014 and Battle Glacier 1400 m from its junction with Thiel Glacier and 3500 m from the Gilkey Glacier. Melkonian, Willis, and Pritchard (2014) note the fastest thinning in the Gilkey Glacier system to be near the terminus and in the lower several kilometers of Thiel Glacier.

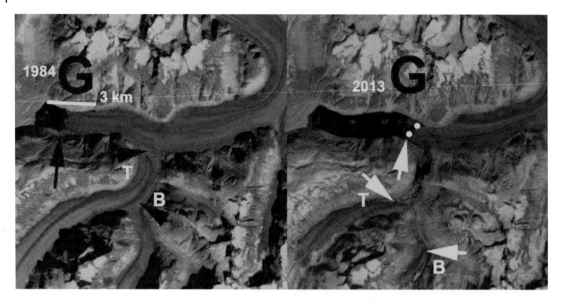

Figure 3.13 Gilkey Glacier change in 1984 and 2013 Landsat images. Yellow and red dots indicate the 2013 and the 1984 terminus positions, respectively. T = Thiel Glacier; B = Battle Glacier. (Landsat, US Geological Survey.)

Arrows #3 and #4 indicate valleys that distributary tongues of the Gilkey Glacier flow into (Fig. 3.16). In 1984, at #3, the glacier extended 1.6 km upvalley ending where the valley split. The portion of the Gilkey flowing into the valley had a medial moraine in its center. At arrow #4, the glacier extended 1.5 km up the Avalanche Canyon. In 2014, at #3, the glacier tongue ends 1.2 km from the valley split, and the medial moraine does not enter the valley. At #4, the glacier has retreated 1.3 km, leaving this valley nearly devoid of a glacier.

Further upglacier, arrow #5 indicates a side glacier that in 1984 featured an unending system of glacier flowing down the steep mountainsides into the valley bottom (Fig. 3.17). In 2014 two rock ribs extended along most of the east and west valley walls separating the glaciers on mountainside from the main valley glacier, which has as a result been reduced in width and velocity. At arrow #6, a tributary glacier is seen merging with Gilkey Glacier in 1984. In 2014 this tributary no longer reached the Gilkey Glacier, ending 300 m up the valley wall. At arrow #7, the Little Vaughan Lewis Icefall in 1984 is seen merging with the Gilkey Glacier across a 300 m wide front. This I can attest from seeing the glacier that summer to be an accurate observation. In 2014 at arrow #7, the Little Vaughan Lewis Icefall no longer feeds ice directly to the Gilkey Glacier. There is still avalanching but not a direct flow connection. The main Vaughan Glacier Icefall is still impressive, just south of the rib beyond arrow #7. Measurements of snowpack are made annually by JIRP above the icefall and indicate a mean snow depth exceeding 3 m in early August. Pelto, Kavanaugh, and McNeil (2013) summarize the results of this ongoing research that Chris McNeil (USGS) is working on to enhance with newer technology.

3.7 Antler Glacier

The Antler Glacier is an outlet glacier of the Juneau Icefield. It is actually a distributary glacier of the Bucher Glacier. It splits from the Bucher Glacier 8.5 km above where the Bucher Glacier joins the

Figure 3.14 The 2013 Landsat image of the Gilkey Glacier region indicating key locations: 1 = Gilkey Lake; 2 = valley of Thiel and Battle Glaciers; 3 and 4 = distributary tongues of the glacier; 5 = tributary glacier; 6 = unnamed glacier separating from Gilkey; 7 = Little Vaughan Lewis Glacier. (Landsat, US Geological Survey.)

Gilkey Glacier as a tributary. In 1948 it spilled over the lip of the Antler River valley from the Bucher Glacier and flowed 6 km downvalley to end in a proglacial lake (Fig. 3.18). The glacier was 6200 m long in 1948; red arrow indicates 1984 terminus and yellow arrow indicates 2014 terminus. In 1984 Antler Glacier no longer reached Antler Lake, which had expanded from a length of 1.6 km in 1948 to 4.2 km. The glacier was still 2.7 km long. In 1997 the lower 2 km of the Antler Glacier was gone, and the glacier ended near the base of the steep eastern entrance to the valley. In 2013 Antler Glacier extended only 400–500 m over the lip of the valley entrance from Bucher Glacier, having lost 2200 m of its length since 1984 and 5.8 km since 1948.

The lake is gorgeous, and the valley once filled by the glacier is now nearly devoid of glacier input. The retreat is largely a result of reduced flow from the thinning Bucher Glacier, which no longer spills over the valley lip significantly. As the Bucher Glacier continues to thin, the Antler Glacier will cease to exist. This thinning is due to increased ablation of the glacier. The mass balance loss at nearby

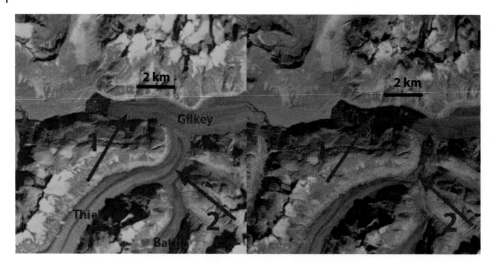

Figure 3.15 Comparison of Gilkey Glacier terminus area with Landsat imagery, 1984–2014. (Landsat, US Geological Survey.)

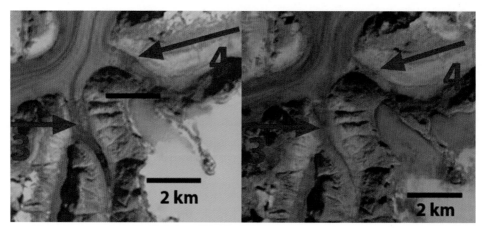

Figure 3.16 Comparison of the Avalanche Canyon area, 1984–2014. (Landsat, US Geological Survey.)

Lemon Creek Glacier from 1953 to 2011 was −26.6 m (Pelto, Kavanaugh, and McNeil, 2013) which equals a thinning of at least 29 m. Gilkey Glacier, which is fed by Bucher Glacier, has retreated 3.2 km from 1984 to 2013 and 4 km from 1948 to 2013 (Pelto *et al.*, 2013). Melkonian, Willis, and Pritchard (2014) indicate thinning of 1–2 m per year for Bucher Glacier from 2000 to 2009.

3.8 Field Glacier

Field Glacier in 1984 ended at the edge of an outwash plain with a few glimpses of a lake developing near its margin (Fig. 3.19). In 2013 a substantial lake has formed at the terminus, and the glacier has retreated 2300 m. A lake has also developed at the first terminus joining from the east, and most of the

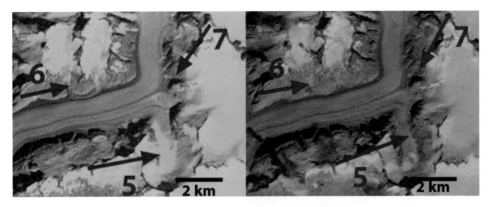

Figure 3.17 Comparison of the Vaughan Lewis Glacier area, 1984–2014.

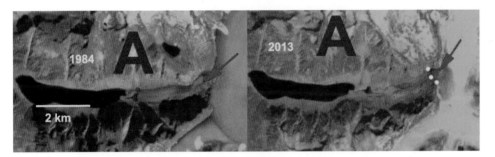

Figure 3.18 Antler Glacier change in 1984 and 2013 Landsat images. Yellow and red dots indicate the 2013 and the 1984 terminus positions, respectively. (Landsat, US Geological Survey.)

width of this glacier has been lost. It is clear that the two lakes will merge as the retreat continues. The Field Glacier in 2006 had developed a proglacial lake at the terminus that averaged 1.6 km in length, with the east side being longer. In 2009 the lake had expanded to 2.0 km long and was beginning to incorporate the incipient lake on the west side of the main glacier tongue. In 2011 the main lake has nearly reached the southern tributary lakes. The lake has expanded to 2.6 km long, with the west side having caught up with the east side and an area of 4.0 km². In addition the main lake has joined with the fringing lake on the south side of the south tributary. There is also a lake on the north side of this tributary. This lake should soon fill the valley of the south tributary and fully merge with the main, as yet unnamed, lake at the terminus; maybe this should be named Field Lake.

3.9 Llewellyn Glacier

The second largest glacier of the icefield is the Llewellyn Glacier, which is in British Columbia. The glacier has several termini; here we examine three of them that have retreated 900 m from 1984 to 2013, at Points A, B, and D (Fig. 3.20). In 1984 the glacier terminated in a shallow, narrow portion of a lake at Point A. At Point B the terminus ended in a deeper, wider proglacial lake. At Point D the glacier ended at a series of small lakes. The snowline during the 1998–2013 period averaged 1900 m, too high for an equilibrium glacier mass balance. This led to substantial thinning. Melkonian, Willis,

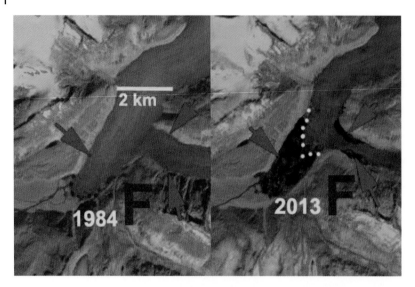

Figure 3.19 Field Glacier change in 1984 and 2013 Landsat images. Yellow and red dots indicate the 2013 and the 1984 terminus positions, respectively. Pink arrows indicate the same three locations. (Landsat, US Geological Survey.)

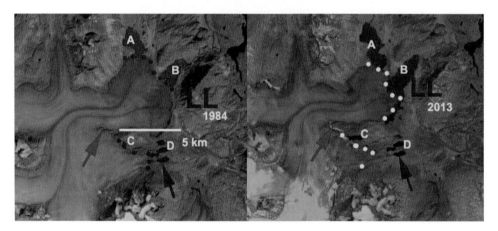

Figure 3.20 Llewellyn Glacier indicating locations at three different terminus tongues in 1984 and 2013 Landsat images. Yellow and red dots indicate the 2013 and the 1984 terminus positions, respectively. (Landsat, US Geological Survey.)

and Pritchard (2014) note thinning of more than 1 m per year at the terminus diminishing to little change above 1500 m from 2000 to 2009. This will drive continued retreat, supplemented by calving into the still growing proglacial lake at Points A and B.

3.10   Tulsequah Glacier

Tulsequah Glacier in 1984 ended at an outwash plain with a small marginal lake beginning to develop (red arrow; Fig. 3.21). In 2013 a large proglacial lake has developed due to the 2500 m retreat. A side

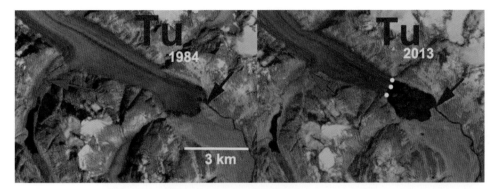

Figure 3.21 Tulsequah Glacier change in 1984 and 2013 Landsat images. Yellow and red dots indicate the 2013 and the 1984 terminus positions, respectively. Pink arrow indicates former glacier-dammed lake location, and red arrow marks the outlet stream at the 1984 terminus location. (Landsat, US Geological Survey.)

valley down which a distributary tongue of the glacier flowed in 1984 has retreated out of the valley in 2013 (pink arrow). The issue driving the retreat is that the equilibrium line where melting equals accumulation and bare glacier ice is exposed has risen and is now typically at 1400 m.

When water stored behind, on, or under a glacier is released rapidly, this outburst is referred to as a jökulhlaup. These outburst floods can pose a serious threat to life and property, but not from the modest floods of the Tulsequah system along this relatively undeveloped watershed. Tulsequah Glacier has a long history of jökulhlaups since the early twentieth century Marcus (1960). The floods resulted after decades of downwasting and retreat of Tulsequah Glacier. In particular a tributary glacier feeding the Tulsequah retreated and downwasted faster than the main glacier. This valley then was dammed by the main stem of the glacier. There is no surface drainage evident from either Lake No Lake or Tulsequah Lake, indicating all discharge is through a subglacial tunnel under the main stem of the glacier emerges at the terminus and causes modest downstream flooding (Fig. 3.22).

3.11 Twin Glacier

The East and West Twin Glaciers are receding up separate fjords, though they are fed from a joint accumulation zone. The East Twin is a narrower glacier and drops more quickly in elevation. The glacier has retreated 900 m from 1984 to 2013 and is now at narrow point in the fjord that appears to be at the fjord head (Fig. 3.23). In 2015 the glacier appeared to have nearly lost contact with Twin Lake.

The West Twin has retreated 600 m to an elbow in the fjord. Elbows like this are often good pinning points that are a more stable setting, temporarily reducing the retreat rate. The terminus has calved into Twin Lake for over a century, but in 2015 the width of the terminus calving into the lake has declined to 150 m from 600 m in 1984. The bedrock exposed on either side of the terminus indicates that the terminus is on the verge of retreating from the lake, which is evident in the 2006 Google Earth imagery (Fig. 3.24). The black arrows indicate both bedrock at the glacier front and the trimlines left from recent thinning. The glacier will no longer be calving, which should also slow the retreat rate. Ogives are apparent on both glaciers generated by the seasonal variation of flow through the icefall that connects the accumulation zone and the ablation zone.

In 2015 the snowline on Twin Glacier was particularly high; the accumulation zone usually covers the entire reach of the broad high-elevation accumulation zone. The declining mass balance identified

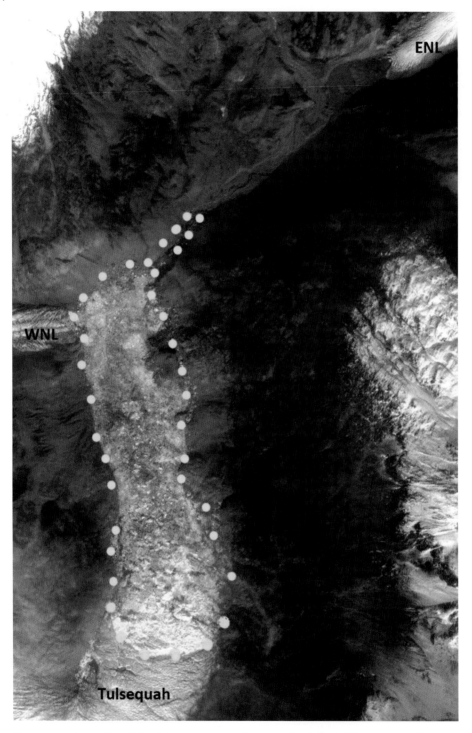

Figure 3.22 A 2010 Google Earth image indicating location of a drained glacier-dammed lake on the east side of Tulsequah Glacier. The lake is drained stranding icebergs, and the lake shape is outlined with yellow dots. (Google Earth.)

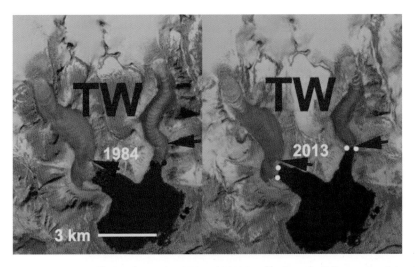

Figure 3.23 Twin Glacier change in 1984 and 2013 Landsat images. Yellow and red dots indicate the 2013 and the 1984 terminus positions, respectively. East and West Twin Glaciers share an accumulation zone and have similar elevation profiles.

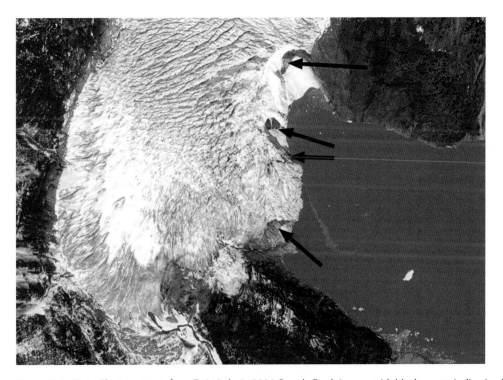

Figure 3.24 Twin Glacier retreats from Twin Lake in 2006 Google Earth image, with black arrows indicating bedrock being exposed between the terminus and the lake. (Google Earth.)

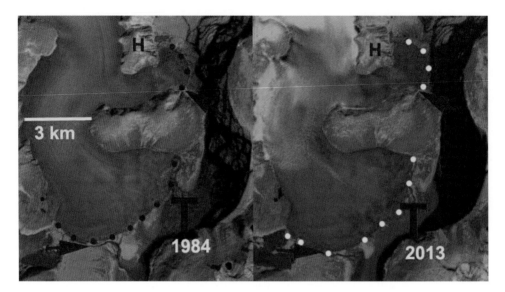

Figure 3.25 Taku Glacier terminus change in 1984 and 2013 Landsat images. Yellow and red dots indicate the 2013 and the 1984 terminus positions, respectively. (Landsat, US Geological Survey.)

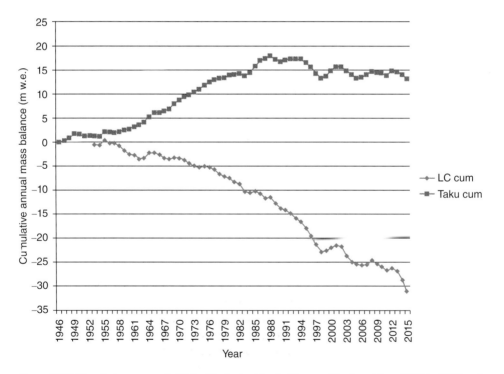

Figure 3.26 Mass balance record of Taku Glacier (red line) with a positive trend from 1946 to 1986 and equilibrium since, versus Lemon Creek Glacier (blue line) indicative of the rest of the glaciers with a declining mass balance throughout the period.

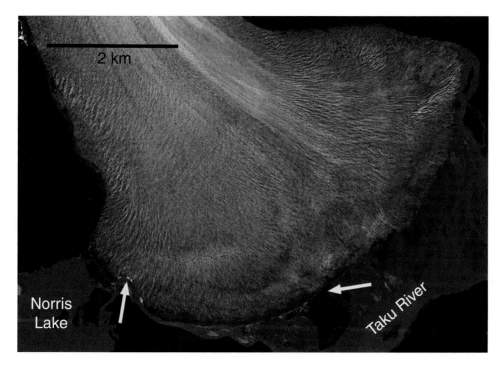

Figure 3.27 Terminus of Taku Glacier in a 2010 Google Earth image. The white arrow at left indicates where the outlet river of Norris Glacier is eroding the Taku terminus. The white arrow at right indicates an island generated by the deformation of outwash plain sediments. (Google Earth.)

by the Juneau Icefield ongoing mass balance program, which the high snowlines are indicative of, is what is driving the retreat (Pelto, Kavanaugh, and McNeil, 2013).

3.12 Taku Glacier

Taku Glacier is the largest glacier of the icefield, and unlike all the others, it has been advancing nonstop over the last century (Fig. 3.25). The sustained positive mass balance from 1946 to 1988 has driven this advance (Pelto, 2011). This led to the glacier thickening along its entire length (Pelto *et al.*, 2008; Larsen *et al.*, 2007). Taku Glacier is noteworthy for its positive mass balance from 1946 to 1988, which resulted from the cessation of calving around 1950 (Pelto and Miller, 1990). The positive mass balance resulting from this dynamic change with calving cessation gives the glacier an unusually high AAR (Fig. 3.26).

Since 1950 observations of velocity near the snowline of the glacier by JIRP indicate that the glacier has had a remarkably steady flow over the past 50 years (Pelto *et al.*, 2008). This occurred during a period of glacier thickening; average profile velocity was 0.5 m d^{-1} (Pelto *et al.*, 2008). Since 1988 the glacier has not been thickening near the snowline as mass balance has declined slightly (Pelto, Kavanaugh, and McNeil, 2013). We have been able to observe the snowline movement in satellite images to help determine the mass balance (Pelto, 2011; Mernild *et al.*, 2013). The changes

Table 3.1 Transient snowline observed on Juneau Icefield glaciers on the same date, with observations from Landsat TM images on the noted dates.

Date	Norris	Taku	Lemon Creek	Mendenhall	Herbert	Eagle
8/29/1995	975	1025	1125	1075	1100	1050
9/4/1996	1050	1075	1150	1125	1150	1175
9/6/1997	1100	1125	1300	1200	1200	1175
9/16/1998	1050	1075	1200	1175	1150	1150
8/31/1999	750	850	950	900	925	900
8/29/2000	750	775	900	875	925	900
8/15/2001	800	800	900	900	925	900
10/3/2002	925	950	1025	1050	1075	1025
10/1/2003	1000	1075	1300	1150	1175	1150
9/1/2004	1075	1050	1100	1150	1200	1200
9/11/2005	1000	1000	1050	1050	1100	1100
9/16/2006	1025	975	1025	1125	1150	1150
9/22/2007	925	925	1000	1050	1050	1025
8/19/2008	700	775	900	850	900	900
9/14/2009	925	950	1050	1050	1050	1050
9/18/2010	950	975	1075	1075	1050	1025
9/11/2011	1000	975	1100	1150	1150	1125
8/31/2012	700	750	925	850	900	850
9/2/2013	950	975	1075	1075	1075	1100
8/20/2014	1025	1000	1150	1100	1125	1175
8/14/2015	1100	1150	1200	1250	1300	1350

The data are the elevation of the snowline in meters (Pelto, Kavanaugh, and McNeil, 2013). The spatial resolution of 30 m, combined with mean surface gradients of $0.04–0.08 \, \mathrm{m \, m^{-1}}$, yields an error of less than $\pm 5 \, \mathrm{m}$ in TSL elevation.

at the glacier front are quite variable as the glacier advances. JIRP measurements of the terminus demonstrate this from 2001 to 2008, with an interactive map from Scott McGee indicating advances in some area, minor retreat in others, and back and forth in some others. There is no change at the east and west sides of the margin from 2008 to 2012; however, there is a $55–115 \, \mathrm{m}$ of advance closer to the center. Motyka and Echelmeyer (2003) in 2001 observed the terminus advancing at $0.3 \, \mathrm{m \, d^{-1}}$ leading to the formation of push moraines up to 200 m beyond the terminus that were up to 7 m high and advancing at $0.1–0.15 \, \mathrm{m \, d^{-1}}$. Motyka *et al.* (2006) found that the glacier base was more than 50 m below sea level within 1 km of the terminus and had deepened substantially since 1984. The current terminus area is within a few meters of sea level (Fig. 3.27). The movement of push moraines and bulges beyond the terminus indicates the glacier front extends below sea level and is shoving the glacial outwash plain, developing the fore bulges.

The advance of this glacier will not end until significant mass balance loss, reduces flow into and thins the expanding terminus lobe. The recent decrease in mean annual balance indicates that the terminus advance will continue to diminish, but retreat is not imminent. The altitude of the snowline at the end of summer has remained above what would be the ELA0 altitude on all Juneau Icefield glaciers in recent years (Table 3.1).

References

Berthier, E., Schiefer, E., Clarke, G.K.C. *et al.* (2010) Contribution of Alaskan glaciers to sea-level rise derived from satellite imagery. *Nature Geoscience*, **3** (2), 92–95. doi: 10.1038/ngeo737

Boyce, E., Motyka, R., and Truffer, M. (2007) Flotation and retreat of a lake-calving terminus, Mendenhall Glacier, southeast Alaska, USA. *Journal of Glaciology*, **53**, 211–224.

Heusser, C.E. and Marcus, M.G. (1964) Surface movement, hydrological change and equilibrium flow on Lemon Creek Glacier, Alaska. *Journal of Glaciology*, **5** (37), 61–75.

Larsen, C.F., Motyka, R.J., Arendt, A.A. *et al.* (2007) Glacier changes in southeast Alaska and northwest British Columbia and contribution to sea level rise. *Journal of Geophysical Research*, **112**, F01007. doi: 10.1029/2006JF000586

Marcus, M.G. (1960) Periodic drainage of glacier-dammed Tulsequah Lake. *British Columbia: Geographical Review*, **50**, 89–106.

Melkonian, A., Willis, M., and Pritchard, M. (2014) Satellite-derived volume loss rates and glacier speeds for the Juneau Icefield, Alaska. *Journal of Glaciology*, **60** (222), 743–760. doi: 10.3189/2014JoG13J181

Mernild, S., Pelto, M., Malmros, J. *et al.* (2013) Identification of snow ablation rate, ELA, AAR and net mass balance using transient snowline variations on two Arctic glaciers. *Journal of Glaciology*, **59**, 649–659. doi: 10.3189/2013JoG12J221

Miller, M.M. and Pelto, M.S. (1999) Mass balance measurements on the Lemon Creek Glacier, Juneau Icefield, AK 1953–1998. *Geografiska Annaler*, **81A**, 671–681.

Motyka, R.J. and Echelmeyer, K.A. (2003) Taku Glacier (Alaska, U.S.A.) on the move again: active deformation of proglacial sediments. *Journal of Glaciology*, **49**, 164.

Motyka, R.J., Truffer, M., Kuriger, E.M., and Bucki, A.K. (2006) Rapid erosion of soft sediments by tidewater glacier advance: Taku Glacier, Alaska, USA. *Geophysical Research Letters*, **33**, 1–5. doi: 10.1029/2006GL028467

Pelto, M.S. (2011) Utility of late summer transient snowline migration rate on Taku Glacier, Alaska. *The Cryosphere*, **5**, 1127–1133.

Pelto, M., Kavanaugh, J., and McNeil, C. (2013) Juneau icefield mass balance program 1946–2011. *Earth System Science Data*, **5**, 319–330. doi: 10.5194/essd-5-319-2013

Pelto, M. and Miller, M.M. (1990) Mass balance of the Taku Glacier, Alaska from 1946 to 1986. *Northwest Science*, **64** (3), 121–130.

Pelto, M.S., Miller, M.M., Adema, G.W. *et al.* (2008) The equilibrium flow and mass balance of the Taku Glacier, Alaska 1950–2006. *The Cryosphere*, **2**, 147–157.

4

Northern Patagonia Icefield Region

Overview

The Andes Mountains in the Patagonia region of Chile and Argentina are host to three significant icefields and numerous regions of mountain glaciers. Davies and Glasser (2012) observed that from 1870 to 2011, 90.2% of glaciers shrank between the end of the Little Ice Age and 2011, 0.3% advanced, and no change was observed for 9%. They further noted that annual rates of shrinkage across the Patagonian Andes increased in each time segment they analyzed (1870–1986, 1986–2001, 2001–2011), with annual rates of shrinkage twice as rapid from 2001 to 2011 as from 1870 to 1986. The rate of change was observed to be fastest for the more northerly glaciers, with the Northern Patagonia Icefield (NPI) glaciers shrinking particularly rapidly. It has been estimated that the wastage of the two icefields from 1995 to 2000 has contributed to sea level rise by 0.105 ± 0.011 mm per year, which is double the ice loss calculated for 1975–2000 (Rignot, Rivera, and Casassa, 2003). Nearly 90% of the glaciers studied are affected. Of the 72 glaciers surveyed in the region, 63 have retreated significantly, only 8 have remained stable, and 1 advanced. Davies and Glasser (2012) identified two periods of fastest recession since 1870: 1975–1986 and 2001–2011 for NPI glaciers. The loss was 0.07% from 1870 to 1986, 0.14% a^{-1} from 1986 to 2001, and 0.22% a^{-1} from 2001 to 2011. The retreat has led to lake expansion. Loriaux and Casassa (2013) examined the expansion of lakes of the Northern Patagonian Ice Cap. From 1945 to 2011 lake area expanded by 65%, or 66 km². The increase in the number of lakes increased the retreat of glaciers terminating in them via calving. The reduction is not restricted to the terminus area. Willis *et al.* (2011) observed that the thinning rate of NPI glaciers below the equilibrium line had increased substantially from 2000 to 2012. Paul and Mölg (2014) observed a more rapid retreat of 25% of the total area lost from glaciers in northern Patagonia from 1985 to 2011; the study area was north of the NPI. The spatial extent of debris cover on the surface of NPI between 1987 and 2015 has increased from 168 km² in 1987 to 307 km² in 2015 (Glasser *et al.*, 2016). The average TSL during the 2013–2016 period is 100 m higher than for the 1979–2003 period (Glasser *et al.*, 2016).

Here we examined the change in these glaciers between a Landsat image from February 9, 1987 (LT52320921987040XXX03 and LT52320931987040XXX02), and January 21, 2015 (LC82320922015021LGN00 and LC82320932015021LGN00); it takes two images to cover the icefield images. There is one small section of San Quintín Glacier that is cloud covered on January 21, 2015, and an image from March 26, 2015, is used (LC82320932015085LGN00). The glaciers will be discussed in a counterclockwise from the north side (Fig. 4.1). Rivera *et al.* (2007) identified the icefield with an area of 3953 km², with much of the icefield a plateau accumulation zone between 1100 and 1600 m. This is an area of extreme precipitation (4–10 m) (Schaefer *et al.*, 2015). San Rafael Glacier is the only glacier that reaches tidewater. In 1987, 7 of the 24 main glacier termini ended

Figure 4.1 Landsat image of the Northern Patagonia Icefield with glaciers. R = Reicher, G = Gualas, SR = San Rafael, SQ = San Quintín, B = Benito, H1 = HPN1, A = Acodado, ST = Steffen, H4 = HPN4, PS = Pared Sud, PN = Pared Nord, C = Colonia, Ca = Cachet, N = Nef, S = Soler, L = Leones, F = Fiero, EX = Exploradores, G = Grosse, and V = Verde. Landsat image is from January 21, 2015, except the terminus of San Quintín, which is from March 26, 2015. (Landsat, US Geological Survey.)

in a lake. Most of the remaining glaciers currently terminate in proglacial lakes, though this was not the case in 1987. In 2015, 19 of the 24 glaciers ended in a lake. As glacier retreat has led to lake development and expansion, calving has been an important factor in increasing retreat rates.

4.1 Reichert Glacier

Reichert Glacier (Reicher) is at the northwest corner of the NPI and flows west from the Mont Saint Valentin region and ends in the expanding Reicher Lake. Rivera *et al.* (2007) noted that the glacier was named after the French geologist Federico Reichert, but Reicher has ended up as the established spelling. They further noted that the glacier lost 4.2 km^2 of area from 1979 to 2001 and had an ELA of 1330 m. The glacier has two main icefalls: one at the first bend in the glacier above the terminus at 400 m and the other at the ELA from 1100 to 1600 m (Fig. 4.2). Davies and Glasser (2012) identified the most rapid area loss of −0.77% per year in the 1986–2001 period. The glacier retreated rapidly from 1987 to 1997, but the terminus was stabilized from 1997 to 2001, before retreating again to the near 2014 terminus in 2002.

Figure 4.2 Landsat comparison of Reichert Glacier in 1987 and 2015. Red arrow indicates the 1987 terminus location, yellow arrow indicates the 2015 terminus location, and purple arrow indicates the upglacier thinning. (Landsat, US Geological Survey.)

Here we examined Landsat imageries from 1987 and 2015 to document the changes. The red and yellow arrows indicate the 1987 terminus and the 2015 terminus, respectively. In 1987 the glacier terminated close to the southern end of Reicher Lake (red arrow). In 1998 the glacier had retreated and terminated on the west side of Reicher Lake across the lake from the main glacier valley. In 2014 the glacier had retreated into the main glacier valley, and Reicher Lake extended 8.8 km from

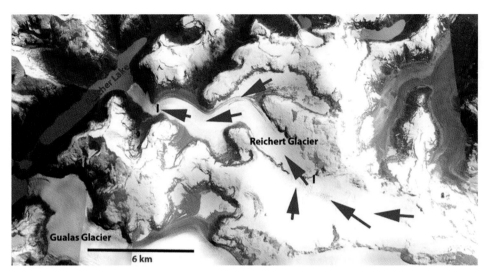

Figure 4.3 A 2013 Google Earth image, indicating flow and two icefalls (I). (Google Earth.)

the northeast to southwest. A small lake is observable above the lowest icefall. This lake indicates a potential second lake basin beginning to develop in the glacier reach above the first icefall (Fig. 4.3). If this is the case, another rapid retreat will ensue, though not in the immediate future. The glacier retreated 6.7 km from 1987 to 2015, with 90% of the retreat occurring in 2002. The area extent loss is 8–9 km^2. The lower icefall is 1.5 km from the current terminus and indicates the maximum extent of Reicher Lake and the retreat that can be enhanced by calving into that lake.

4.2 Gualas Glacier

Gualas Glacier drains from the northwest portion of the NPI into a rapidly expanding new lake, Lago Gualas. There is a spectacular icefall (I) where the glacier descends from the main NPI from 1600 to 900 m, which is below the equilibrium line (Fig. 4.4). Below this point, the Gualas has a 2 km wide, 15 km long valley reach extending to the terminus which currently has a substantial calving face into Lago Gualas (A–B). Point M indicates an area of substantial moraine cover on the ice, which will with continued retreat be a likely location for a new lake to form, much as the periodic lake at Point C (Fig. 4.5).

Lopez *et al.* (2010) have documented a 1.8 km retreat from 2001 to 2011 of this glacier, updating and expanding on the work of Rivera *et al.* (2007). They have further noted an average thinning of the valley tongue of 2.1 m per year from 1975 to 2005 and a doubling in the rate of area lost (2.8 km^2) during the period 2001–2011 versus 1975–2001. In 1987 the glacier essentially filled the lake basin. In 2001 an evident fringe of water separated the glacier from the lake margin on all but the eastern side of the lake. In 2005 the margin was hard to discern given the extensive floating icebergs in the lake. In 2011 the lake was evident and the glacier had retreated 2.2 km from its 1987 position. In 2015 the retreat from 1987 is 3 km.

The glacier surface is steep in the first kilometer behind the terminus, which indicates rising bedrock under the glacier. Then the glacier has a very modest slope for the next 14 km. As long as the glacier can calve into a lake, this will enhance retreat. The current lake may not end at the bedrock

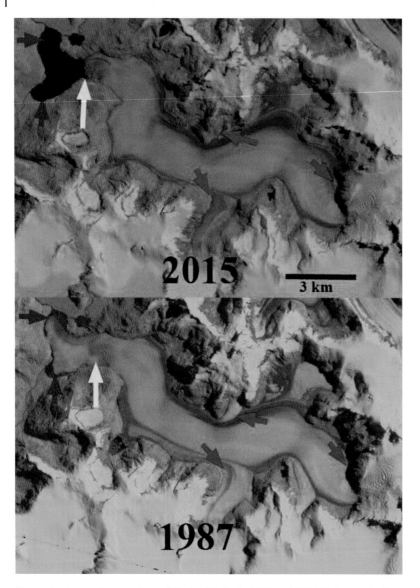

Figure 4.4 Landsat comparison of Gualas Glacier in 1987 and 2015. Red arrow indicates the 1987 terminus location, yellow arrow indicates the 2015 terminus location, and purple arrow indicates the upglacier thinning. (Landsat, US Geological Survey.)

step just behind the current terminus. However, even if this marks the end of the lake, the low slope above that point indicates another basin that will have sufficient depth to form a second lake basin. It is unlikely that the calving retreat of this glacier will have more than a temporary interruption. Of particular interest is the 2.1-m per year thinning that has occurred on the lower glacier, while on the upper glacier there is a small amount of thickening (Rivera *et al.*, 2007). This implies that the retreat is driven by enhanced melting due to warming, since the only way to thicken the glacier in the accumulation zone is via increased snowfall.

Figure 4.5 A 2014 Landsat image. Point M indicates an area of substantial moraine cover on the ice, Point C indicates a periodic lake, and Point I indicates the key Icefall with the snowline at the top. (Landsat, US Geological Survey.)

4.3 San Rafael Glacier

The lowest latitude tidewater glacier, San Rafael Glacier is noteworthy for its high velocities and calving rate (Fig. 4.6) (Warren et al., 1995). Warren and Aniya (1999) identified a calving rate of $4500\,\mathrm{m\,a^{-1}}$, while Mouginot and Rignot (2015) noted a peak glacier velocity of $7200\,\mathrm{m\,a^{-1}}$. Rivera *et al.* (2007) identified the glacier area in 2001 to be $722\,\mathrm{km^2}$, with an ELA at 1015 m. Given the water depth of 140 m, this is an unusually high rate compared to the general relation between water depth and calving rate (Brown, Meier, and Post, 1982; Pelto and Warren, 1991). The high retreat rate may in part be related to the high water temperatures in the lagoon (Luckman *et al.*, 2015). The region of high velocity extends 20 km inland. In the 1980s, the terminus retreated at $300\,\mathrm{m\,a^{-1}}$, but did not retreat from 1990 to 1992 (Warren and Aniya, 1999). From 1987 to 2015, the glacier had retreated 1.3 km. Upglacier there is significant thinning evident at the three purple arrows. Willis *et al.* (2012) noted a thinning rate in the ablation zone of $2.6\,\mathrm{m\,a^{-1}}$ from 2000 to 2011 and a longer term thinning rate of $2\,\mathrm{m\,a^{-1}}$ since 1945. This suggests that the decline in volume flux will in turn increase the retreat rate. The thinning will also lead to the development of expanded subsidiary calving areas; this will precondition the glacier for enhanced retreat.

4.4 San Quintín Glacier

San Quintín Glacier is the largest glacier of the NPI at $790\,\mathrm{km^2}$ in 2001 (Rivera *et al.*, 2007). In 1987 it was a piedmont lobe with evident marginal proglacial lake development (Fig. 4.7). The peak velocity is $1.1\,\mathrm{km\,a^{-1}}$ near the ELA (Rivera *et al.*, 2007), declining below $1\,\mathrm{m\,d^{-1}}$ in the terminus region. The velocity at the terminus had increased from 1987 to 2014 as the glacier had retreated into the proglacial lake (Mouginot and Rignot, 2015). The high-velocity zone extends more than 40 km inland,

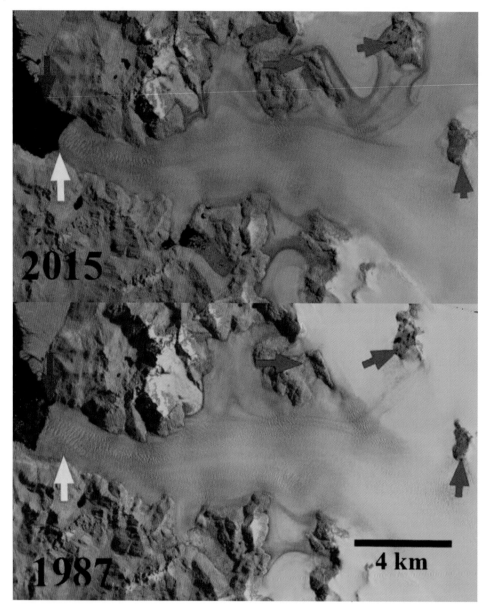

Figure 4.6 Landsat comparison of San Rafael Glacier in 1987 and 2015. Red arrow indicates the 1987 terminus location, yellow arrow indicates the 2015 terminus location, and purple arrow indicates the upglacier thinning. (Landsat, US Geological Survey.)

an even greater distance than at San Rafael (Mouginot and Rignot, 2015). The thinning rate in the ablation zone of the glacier is 2.3 m a^{-1} (Willis *et al.*, 2012). This is leading to the retreat not just of main terminus but also of the distributary terminus areas extending north and south into lake basins from the main glacier. From 1987 to 2015, the main terminus retreated 2200 m. The lake developing on the north side of the glacier is now over 8 km^2.

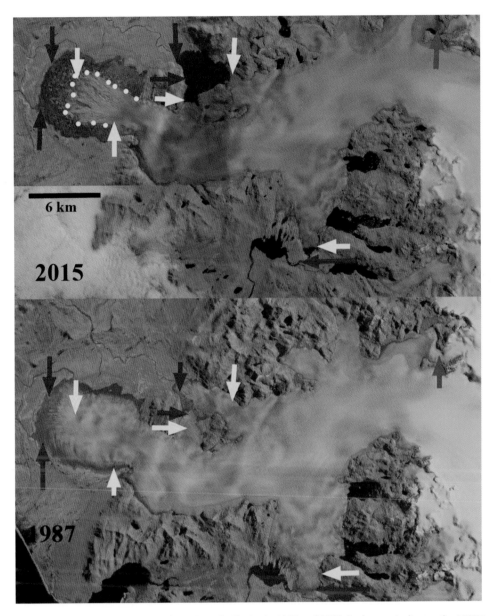

Figure 4.7 Landsat comparison of San Quintín Glacier in 1987 and 2015. Red arrow indicates the 1987 terminus location, yellow arrow indicates the 2015 terminus location, and purple arrow indicates the upglacier thinning. (Landsat, US Geological Survey.)

4.5   Fraenkel Glacier

Fraenkel Glacier drains the west side of the Northern Patagonia Ice Cap just south of San Quintín Glacier. The retreat of this glacier for the last 30 years mirrors that of Gualas and Reichert Glaciers, which also terminate in an expanding proglacial lake (Fig. 4.8). Willis *et al.* (2011) observed in Fraenkel

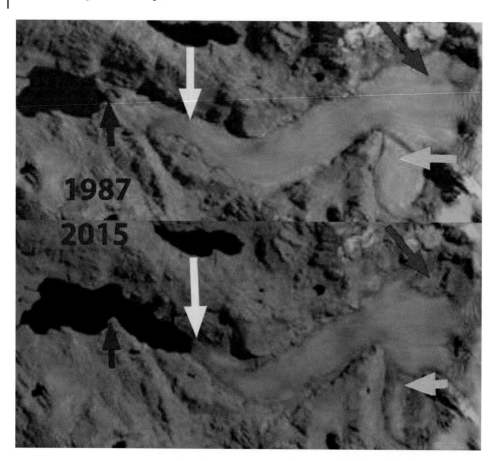

Figure 4.8 Landsat comparison of Fraenkel Glacier in 1987 and 2015. Red arrow indicates the 1987 terminus location, yellow arrow indicates the 2015 terminus location, and purple arrow indicates the upglacier thinning. (Landsat, US Geological Survey.)

Glacier a $2.4\,\text{m}\,\text{a}^{-1}$ thinning in the ablation zone from 2000 to 2010. Mouginot and Rignot (2015) observed that Fraenkel Glacier does not have the high velocity of the neighboring Benito and San Quintín Glacier which leads to the potential for greater mass loss of the ablation zone and even faster retreat. In 1987 the glacier terminus was at the end of a peninsula (red arrow), and the proglacial lake it terminates in was 2 km long. The terminus retreated 800 m by 2000. In 2015 the area around the purple arrow had been deglaciated, emphasizing the amount of thinning in the ablation zone even well upglacier of the terminus. The main terminus in 2015 was at the yellow arrow, indicating a retreat of 1.4 km since 1987. The retreat rate of $50\,\text{m}\,\text{a}^{-1}$ per year, though large, is less than on Reichert or Gualas Glaciers. The glacier valley expands in width above the current terminus, which also would tend to increase retreat.

4.6 Benito Glacier

Benito Glacier in 1987 terminated on an outwash plain. The glacier has five key distributary termini, two of which have open proglacial lakes. In 2015 there were six tributary termini, with five ending

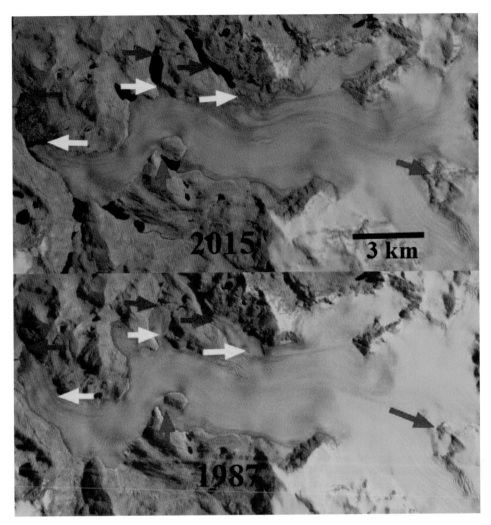

Figure 4.9 Landsat comparison of Benito Glacier in 1987 and 2015. Red arrow indicates the 1987 terminus location, yellow arrow indicates the 2015 terminus location, and purple arrow indicates the upglacier thinning. (Landsat, US Geological Survey.)

in lakes and one having retreated out of a lake basin. The two tributaries to the north indicated with arrows each retreated close to 1 km and in both cases were no longer calving termini. The main glacier terminus had retreated into a proglacial lake, with a retreat of 2 km from 1987 to 2015 (Fig. 4.9). Winchester *et al.* (2014) identified thinning of 150 m in the lower ablation zone from 1973 to 2011, with the most rapid thinning from 2007 to 2011.

4.7 Acodado Glacier

Rio Acodado has two large glacier termini at its headwater, HPN2 and HPN3, that are fed by the same accumulation zone and comprise the Acodado Glacier (Figs. 4.10 and 4.11). The glacier separates

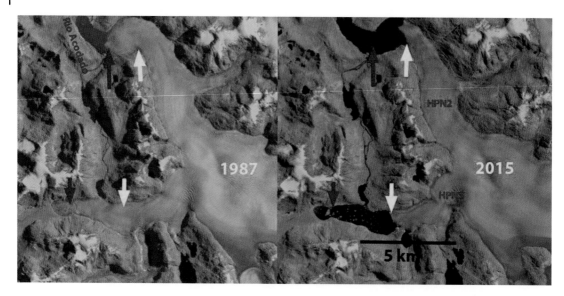

Figure 4.10 Landsat comparison of Acodado Glacier in 1987 and 2015. Red arrow indicates the 1987 terminus location, yellow arrow indicates the 2015 terminus location, and purple arrow indicates the upglacier thinning. (Landsat, US Geological Survey.)

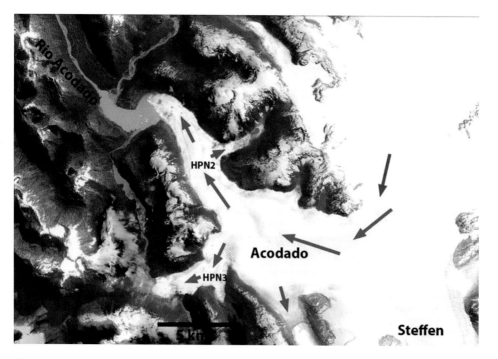

Figure 4.11 Digital Globe image of Acodado Glacier and the termini HPN2 and HPN3.

from Steffen Glacier at 900 m, draining west from the NPI. The lakes at the terminus of each were first observed in 1976 and had an area of 2.4 and 5.0 km^2 in 2011 (Loriaux and Casassa, 2013). Davies and Glasser (2012) noted that the Acodado Glacier termini, HPN2 and HPN3, had retreated at a steadily increasing rate from 1870 to 2011. Willis *et al.* (2012) noted a 3.5 m a^{-1} loss from 2001 to 2011 in the ablation zone of the Acodado Glacier. They also noted that the annual velocity is less than 300 m a^{-1} in the ablation zone.

HPN2 and HPN3 terminated at the red arrow in 1987 and the snowline is at 1000 m. In 2000 the glacier had retreated from the red arrows by 400 m, and the snowline is at 1100 m. In 2014 there were many large icebergs in the lake at the terminus of HPN3, which are from recent calving retreat. This is not an area where the lakes develop even seasonal lake ice cover. The snowline is again at 1100 m. In 2015 it was apparent that HPN2 had retreated 2.1 km since 1987, from the red arrow to the yellow arrow. HPN3 had retreated 3.3 km from the yellow to the red arrow. The retreat accelerated after 2000 for both glaciers. This height of a snowline indicates warm temperatures generating high ablation rates, which will lead to more retreat. HPN3 has a sharp rise in elevation of 2.5 km above the terminus; before it joins the main Acodado Glacier, it should retreat rapidly toward this point and then calving will end and retreat will slow.

4.8 Steffen Glacier

Steffen Glacier is the main south flowing glacier from NPI (Fig. 4.12). Several key research papers have reported on the spectacular retreat of this glacier in recent years. Rivera *et al.* (2007) reported that Steffen Glacier lost 12 km^2 and had an average thinning of 1.5 m in the ablation zone from 1979 to 2001. The JAXA EORC (2011) report compared the Steffen Glacier terminus change from 1987 to 2010. They noted a retreat of approximately 2.1 km of the main stem and 870 m of a western terminus. They also noted a remarkable collapse of the terminus tongue.

In 1987 the lake at the terminus of the glacier was 1.3 km from north to south. The northwestern most terminus is 4 km from the main trunk of Steffen. The middle west terminus (yellow arrow) extends 1.7 km west from the main trunk. There is no lake evident at the red arrow. In 1999 the northwest terminus had retreated little but is showing signs of breakup (Fig. 4.13). The middle west terminus (yellow arrow) had retreated to within a half kilometer of the main trunk in 1999. The main terminus had retreated little since 1987 on the west side but had retreated 700 m on the east side. In 2003, the northwest terminus, a retreat has begun exposing a new section of lake. The middle terminus cannot be discerned. The main terminus is now an isolated tongue in the midst of the terminus lake, with open water along the east and west margins. In 2015 the main terminus had retreated 3.4 km from 1987; there are small lakes on either side of the glacier above the terminus, indicating that the lower 2 km of the glacier is likely to be lost soon (the red and pink arrows). The northwest and middle west termini are both nearly back to the main glacier having retreated 3.8 and 1.3 km, respectively, since 1987.

4.9 HPN4 Glacier

HPN4 Glacier drains the southern side of NPI just east of Steffen Glacier (Fig. 4.14). The terminus had retreated little in this period. The main change is in the eastern tributary 1–2 km north of the terminus. In 1987 there were five separate feeder ice tongues descending from the ice cap into this valley. In 2015 there is just one. Further this tongue has narrowed and downwasted and a new lake is

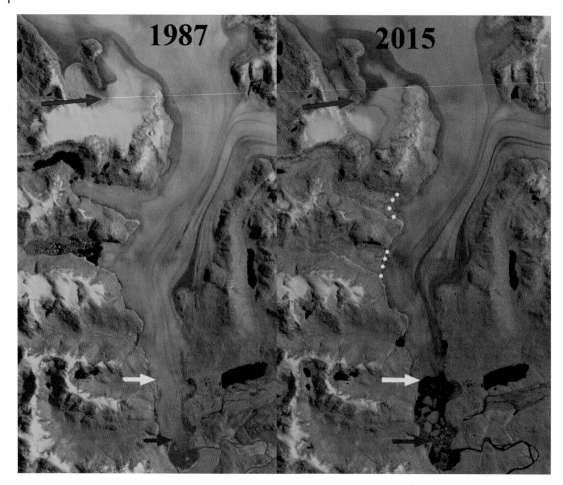

Figure 4.12 Landsat comparison of Steffen Glacier in 1987 and 2015. Red arrow indicates the 1987 terminus location, yellow arrow indicates the 2015 terminus location, and purple arrow indicates the upglacier thinning. (Landsat, US Geological Survey.)

developing. The previous flow diagram indicates the converging flow of the downwasting tributary and HPN4, which meet at the medial moraine. The medial moraine has shifted east as the main HPN4 tongue is beginning to flow up this valley (Fig. 4.15).

In 1987 there were five contributing glacier tongues to the downwasting tributary, each indicated with a red arrow. It is like a bathtub being filled with five taps at once. The yellow arrow indicates a medial moraine at the mouth of the valley, signaling the lack of current contribution of the downwasting tributary to HPN4 Glacier. The medial moraine has shifted east, indicating that the main HPN4 Glacier is now flowing into the valley instead of the downwasting tributary being a contributing tributary to HPN4. In 2015 there was only one contributing glacier tongue to the downwasting tributary, only one tap for this draining bathtub, while the other four contributing tongues have retreated from contact with the downwasting tributary. The medial moraine has spread eastward, and some fringing proglacial/subglacial lakes are evident. The rifts are a sign of instability and typically lead to breakup

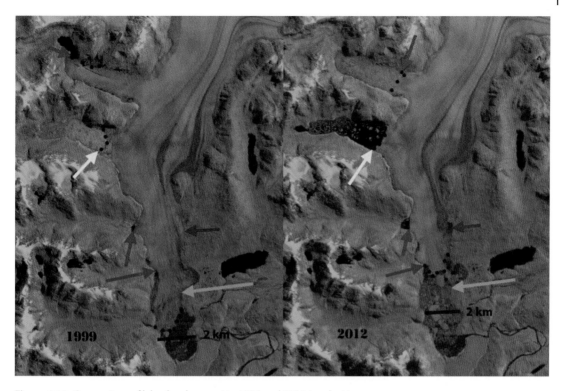

Figure 4.13 Comparison of lake development in 1999 and 2012 Landsat images.

of this portion of the terminus. The downwasting tributary continues to demise faster than HPN4 Glacier, which crosses the valley mouth; hence, it is likely that a glacier-dammed lake will form and that HPN4 Glacier will continue to flow further east up this valley, which could offset some of the downwasting and lake development. In either case this redirected flow of HPN4 into a high ablation valley will help encourage a faster retreat of the main terminus. How large the lake gets and how much of the time it is filled are difficult to speculate upon. Schaefer *et al.* (2013) discussed the HPN4 Glacier because the main terminus has changed little given its modeled mass balance, and the modeled mass balance to the east appears too negative, which they suggest indicates wind redistribution from the HPN4 to the Pared Sud Glacier just east. That is a challenge to sort out without some ground truth.

4.10 Colonia Glacier

Baker River (Rio Baker) is located to the east of the NPI and is fed mainly by glacier meltwater originating from the eastern outlet glaciers of the icefields Leones, Soler, Nef, and Colonia. Rio Baker is the most important Chilean river in terms of runoff, with an annual mean discharge of about $1000 \, \text{m}^3 \, \text{s}^{-1}$. Colonia Glacier drains east from the NPI feeding the Baker River, Chile (Fig. 4.16). It is the largest glacier draining east from the NPI. A comparison of the 1987 and 2015 images indicates a 2.5 km retreat of the glacier front and development of a large lake and areas of thinning well upglacier at the purple arrows. The recent substantial retreat of Colonia Glacier like Nef Glacier just to its north is

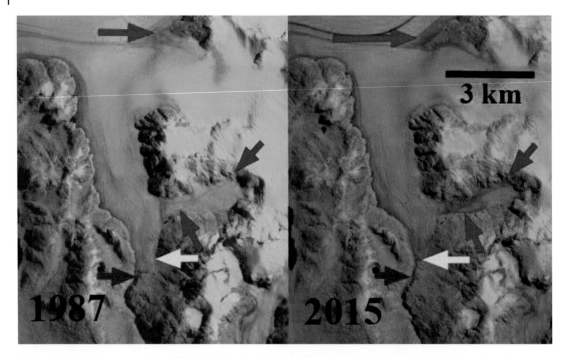

Figure 4.14 Landsat comparison of HPN4 Glacier in 1987 and 2015. Red arrow indicates the 1987 terminus location, yellow arrow indicates the 2015 terminus location, and purple arrow indicates the upglacier thinning.

Figure 4.15 Changes in width, moraine position, and contributing glaciers to the eastern tributary.

posing new hazards. The glacier is unusual in the number of lakes that are adjacent to or feed into the adjacent glacier-dammed or proglacial lakes. In the case of Baker River, the outburst floods are a threat to the planned hydropower developments as documented by Dusaillant *et al.* (2010). The HidroAysén Project proposed five dams on the Baker and Pascua Rivers, generating 2750 MW of power. This river had a series of proposed hydropower projects that have now been canceled by the Chilean government.

The glaciers' recent retreat and glacier lake outburst floods have been closely monitored by the Laboratorio de Glaciología in Valdivia, Chile. Aniya (1999) observed that Colonia Glacier began a rapid retreat after 1985 from 1997 to 2005 that has further accelerated, with a general frontal retreat of 2 km. Rivera *et al.* (2007) observed that the Colonia Glacier had lost 9.1 km^2 of area from 1979 to 2001, which is 3% of the total glacier area and thinned 1.1 m per year in the ablation zone.

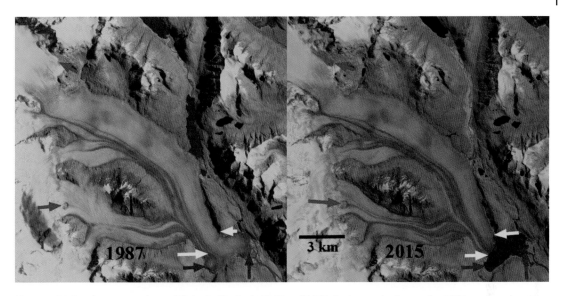

Figure 4.16 Landsat comparison of Colonia Glacier in 1987 and 2015. Red arrow indicates the 1987 terminus location, yellow arrow indicates the 2015 terminus location, and purple arrow indicates the upglacier thinning.

In the spring of 2008, Baker River suddenly tripled in size, and in less than 48 hours, roads, bridges, and farms were severely damaged. Lake Cachet 2 a 5 km^2 glacial lake emptied 200 million m^3 of water in just a matter of hours (Fig. 4.17). This lake drained beneath the glacier after sufficient water had filled the lake to buoy part of the glacier, and subglacial conduits had begun to develop. Since Cachet 2 emptied in April 2008, the lake had emptied at least six more times (in October and December 2008, March and September 2009, March 2010, and early 2013), with peak flow release rate of 3000 m^3 s^{-1} (Dussaillant *et al.*, 2010). A comparison of before and after drainage in Landsat images from September 2012 and February 2013 is shown in Fig. 4.17.

The two lakes at the terminus of the glacier did not exist in 1979; the westernmost terminus lake drained into the easternmost terminus lake via a subglacial tunnel after formation in the late 1980s until 2005 when a channel was cut right through the glacier terminus. The retreat of the glacier terminus first led to significant lake development in 2001. This is evident in the image shown in Fig. 4.18; there is still glacier ice on both sides of this drainage channel. In 2015 the lakes had merged into a single large proglacial lake at the terminus that is 3.2 km wide. The development and demise of glacier-dammed lakes and the resultant problem of glacier lake outburst floods are not rare today.

4.11 Nef Glacier

Nef Glacier began to retreat into a moraine-dammed proglacial lake in 1945 (Loriaux and Casassa, 2013). In 1987 the lake remained less than 1 km long. From 1987 to 2015, the glacier had retreated 1.8 km (Fig. 4.19). In 2015 Nef Glacier had not reached the head of this proglacial lake and will continue to retreat. The west side of the terminus is covered with debris and has a fringing proglacial lake that has developed after 2000 and will aid in the continuing retreat. The lack of elevation change of

Figure 4.17 Landsat images illustrate the impact of a lake-draining event on both the Colonia and Cachet Lakes.

the lower glacier and the isolated proglacial lake suggests that the lake will expand laterally as well as lengthwise. The purple arrows indicate thinning upglacier in a former tributary glacier.

The neighboring Cachet Glacier just west of Nef Glacier had also retreated since 1987. The glacier had retreated 350 m into an eastward-oriented section of Cachet Lake.

4.12   Leones Glacier

Leones Glacier drains east into Lago Leones Lake, with three tributaries joining just above the terminus (Fig. 4.20). Lago Leones feeds the Leones River which is also fed by the retreating Lago General Carrera Glacier. In March 2015, Jill Pelto, my daughter returning from a fieldwork with UMaine in the Falkland Island, took a picture (out the plane window) of Leones Glacier of the NPI (Fig. 4.21). The picture illustrated two changes worthy of further examination. The picture indicates outlet glaciers of the NPI fed by the snow-covered expanse. Also evident is a large landslide (orange arrow) that is fresh, and it showed a new lake that had formed due to retreat of the glacier north of Leones Glacier (red arrow) – hereafter designated North Leones Glacier. The landslide extends 2 km across the glacier and is 3 km from the terminus. Here we used the 1987–2015 Landsat imageries to identify changes in North Leones Glacier and the landslide appearance.

In 1987 there were medial moraines on the glacier surface, but no large landslide deposit. The North Leones Glacier terminated on land (red arrow). A distributary terminus almost connects with another glacier to the north at the yellow arrow. In 2002 a small lake was beginning to form at the terminus of North Leones Glacier. In February 2014 a substantial lake had formed at the end of the North

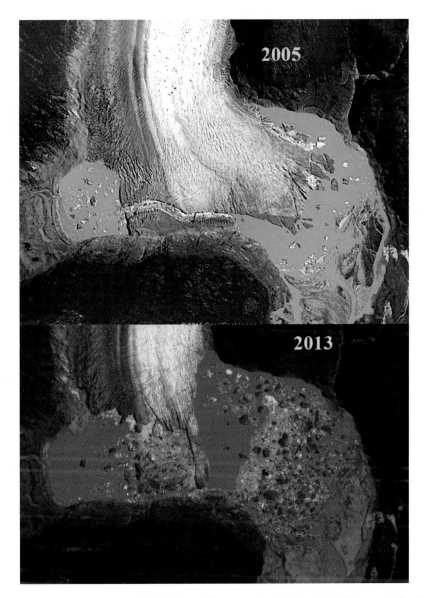

Figure 4.18 Comparison of the Colonia Glacier terminus in Google Earth images indicating the collapse of the terminus tongue and lake development. Google Earth.

Leones Glacier. There is no landslide deposit either. In a January 2015 Landsat image, the landslide deposit was evident, extending about 2 km across Leones Glacier and 3 km from the terminus. The North Leones Glacier had retreated 700 m from 1985 to 2015. The retreat of the distributary terminus indicates thinning upglacier of the icefall on North Leones Glacier. The landslide adds mass to Leones Glacier, which will lead to a velocity increase. The debris is thick enough to reduce melting in this portion of the ablation zone. The velocity of this glacier is indicated by Mouginot and Rignot (2015)

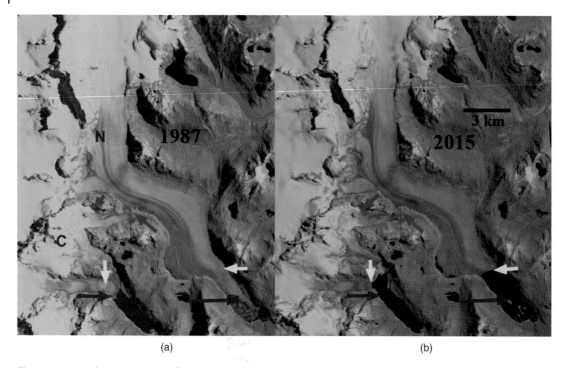

(a)   (b)

Figure 4.19 Landsat comparison of (a) Cachet and (b) Nef Glaciers in 1987 and 2015. Red arrow indicates the 1987 terminus location, yellow arrow indicates the 2015 terminus location, and purple arrow indicates the upglacier thinning.

as 200–400 m per year, indicating that for the next decade at least this landslide will impact the lower Leones Glacier. Willis *et al.* (2012) identified thinning of the Leones Glacier area around 1 m per year, which will be reduced on the landslide arm of the glacier. Leones Glacier had retreated less than 200 m since 1987, the least of any outlet glacier of the NPI.

4.13   Fiero Glacier

Fiero Glacier drains from the northwest region of NPI and is heavily covered with debris in the ablation zone. Fiero Glacier in 1987 ended in a proglacial lake at a peninsula where the lake narrows (red arrows). In 2015 the glacier had retreated 1.5 km into a wider portion of the lake. The lowest 1 km of the glacier is thin and has a low slope and will be quickly lost.

4.14   Grosse Glacier

In Grosse Glacier, the lower 10 km is heavily covered with debris, which typically would slow its response to climate change. Grosse Glacier in 1987 ended on a proglacial outwash plain, with little evident recent retreat (Fig. 4.22), with a few small supraglacial lakes near the margin. In 2015 the glacier had retreated 2.4 km with the commensurate growth of a proglacial terminus lake. The lower 4 km of the glacier appeared relatively stagnant and is poised to be lost as retreat continues.

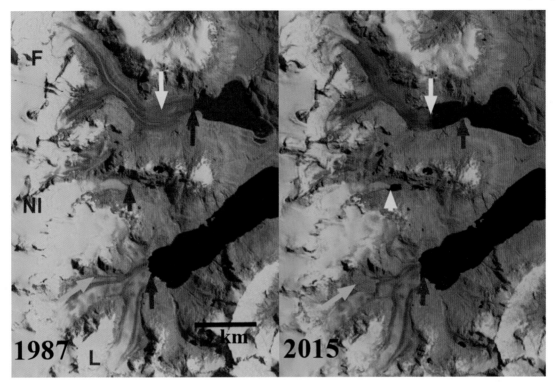

Figure 4.20 Landsat comparison of Leones (L), North Leones (NL), and Fiero (F) Glaciers in 1987 and 2015. Red arrow indicates the 1987 terminus location, yellow arrow indicates the 2015 terminus location, and purple arrow the indicates upglacier thinning.

Thinning is evident at the divide with Verde Glacier (purple arrow), as the ridge between the glaciers has expanded.

4.15 Verde Glacier

Verde Glacier is between Reichert Glacier and Grosse Glacier and drains the north side of NPI. It is a small glacier compared to many outlet glaciers of the icefield. The glacier flows from a pair of peaks at 1800 m to terminate at the edge of a proglacial lake. There is a significant icefall at 800–1250 m and a significant avalanche fan at the base of this icefall that spills from a disconnected portion of the glacier west of the terminus and just northwest of the icefall. Davies and Glasser (2012) indicated the glacier nearly filling the entire lake in 1975. In their Figure 8a they indicated the fastest retreat for the glacier being from 1998 to 2014. Rivera *et al.* (2007) indicated the ELA at the top of the icefall (1250 m). In 1987 the glacier terminated at the red arrow just beyond the northeast bend in the lake. The lake is 1.5 km long (Fig. 4.22). In 2001 the glacier had retreated 400 m and was at this bend. In 2015 the glacier had retreated to the yellow arrow, which is a further 600 m retreat since 2001.

The lake is 2.5 km long measured along its centerline. The 1 km retreat in 28 years is substantial for a glacier that is only 5 km long. The lowest 300 m of the glacier is transitioning from debris-covered ice to an ice-cored moraine and is not active glacier ice. It is certain that the lake will expand further

Figure 4.21 Image of Leones Glacier indicating landslide at orange arrows. Jill Pelto took this picture on March 13, 2015.

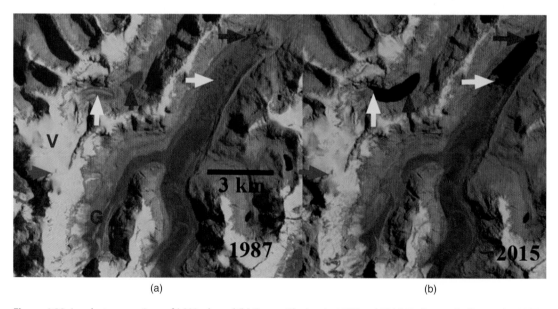

(a)   (b)

Figure 4.22 Landsat comparison of (a) Verde and (b) Grosse Glaciers in 1987 and 2015. Red arrow indicates the 1987 terminus location, yellow arrow indicates the 2015 terminus location, and purple arrow indicates the upglacier thinning.

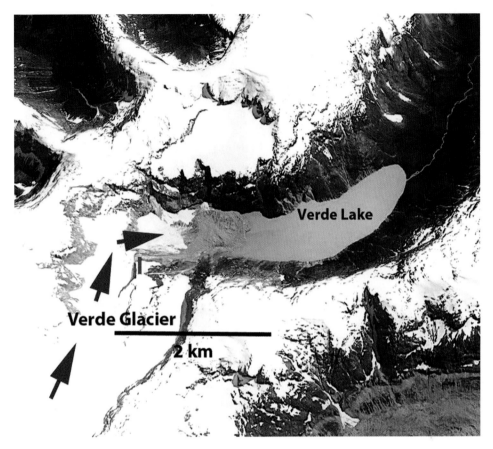

Figure 4.23 Google Earth image indicating flow of Verde Glacier and recession from Verde Lake. (Google Earth.)

as the buried ice melts, but it is nearing its southwestern limit (Fig. 4.23). The steep slope of the icefall and the rock slope to the right of the icefall are evident. This will lead to continued avalanching onto the terminus area, which will make the lowest region more difficult to melt out.

References

Brown, C., Meier, M., and Post, A. (1982) Calving speed of Alaska tidewater glaciers, with application to Columbia Glacier. USGS Prof. Pap. 1258-C.

Davies, B.J. and Glasser, N.F. (2012) Accelerating shrinkage of Patagonian glaciers from the Little Ice Age (AD 1870) to 2011. *Journal of Glaciology*, **58** (212), 1063–1084. doi: 10.3189/2012JoG12J026

Dussaillant, A., Benito, G., Buytaert, W. *et al.* (2010) Repeated glacial-lake outburst floods in Patagonia: an increasing hazard? *Natural Hazards*, **54** (2), 469–481.

Glasser, N., Holt, T., Evans, Z., Davies, B., Pelto, M. and Harrison, S. (2016). Recent spatial and temporal variations in debris cover on Patagonian glaciers. *Geomorphology*, **273**, 202–216.

JAXA EORC (2011) *Significant Retreats of Huge Glaciers in Patagonia, South America*, http://www.eorc .jaxa.jp/en/imgdata/topics/2011/tp110525.html (accessed 15 June 2015).

Lopez, P., Chevallier, P., Favier, V. *et al.* (2010) A regional view of fluctuations in glacier length in southern South America. *Global and Planetary Change*, **71**, 85–108.

Loriaux, T. and Casassa, G. (2013) Evolution of glacial lakes from the Northern Patagonia Icefield and terrestrial water storage in a sea-level rise context. *Global and Planetary Change*, **102**, 33–40.

Luckman, A., Benn, D., Cottier, F. *et al.* (2015) Calving rates at tidewater glaciers vary strongly with ocean temperature. *Nature Communications*, **6**, 8566.

Mouginot, J. and Rignot, E. (2015) Ice motion of the Patagonian Icefields of South America:1984–2014. *Geophysical Research Letters*, **42**. doi: 10.1002/2014GL062661

Paul, F. and Mölg, N. (2014) Hasty retreat of glaciers in northern Patagonia from 1985 to 2011. *Journal of Glaciology*, **60** (224), 1033–1043.

Pelto, M. and Warren, C. (1991) Relationship between tidewater glacier calving velocity and water depth at the calving front. *Annals of Glaciology*, **15**, 115–118.

Rignot, E., Rivera, A., and Casassa, G. (2003) Contribution of the Patagonia Icefields of South America to sea level rise. *Science*, **302**, 434–437.

Rivera, A., Benham, T., Casassa, G. *et al.* (2007) Ice elevation and areal changes of glaciers from the Northern Patagonia icefield, Chile. *Global and Planetary Change*, **59**, 126–137.

Schaefer, M., Machguth, H., Falvey, M. *et al.* (2015) Quantifying mass balance processes on the Southern Patagonia Icefield. *The Cryosphere*, **9**, 25–35. doi: 10.5194/tc-9-25-2015

Warren, C., Glasser, N., Harrison, S. *et al.* (1995) Characteristics of tide-water calving at Glaciar San Rafael, Chile. *Journal of Glaciology*, **41** (273–289), 1995.

Warren, C. and Aniya, M. (1999) The calving glaciers of southern South America. *Global and Planetary Change*, **22**, 59–77.

Willis, M.J., Melkonian, A.K., Pritchard, M.E., and Ramage, J.M. (2012) Ice loss rates at the Northern Patagonian Icefield derived using a decade of satellite remote sensing. *Remote Sensing of Environment*, **117**, 184–198.

Winchester, V., Sessions, M., Cerda, J.V. *et al.* (2014) Post-1850 changes in Glacier Benito, North Patagonian Icefield, Chile. *Geografiska Annaler: Series A*, **96** (1), 43–59.

Schaefer, M., Machguth, H., Falvey, M., and Casassa, G. (2013) Modeling past and future surface mass balance of the Northern Patagonian Icefield. *Journal of Geophysical Research: Earth Surface*, **118**, 571–588. doi: 10.100/jgrf.20038

Aniya, M. (1999) Recent glacier variations of the Hielos Patagónicos, South America, and their contribution to sea-level change. *Arctic, Antarctic and Alpine Research*, **31** (2), 165–173.

5

South Georgia, Kerguelen, and Heard Islands

Overview

South Georgia, Kerguelen, and Heard Islands are all in or just south of the Antarctic Convergence Zone, which is the polar front separating Antarctic and temperate air masses. Surrounded by cold waters originating in Antarctica and in the path of frequent cyclonic storms traveling west along the Antarctic Convergence zone, the islands have a colder climate than expected from its latitude. This leads to considerable glacier cover that has entered a period of rapid and widespread retreat.

South Georgia Island is at 54–55 S and 36–38 W. More than half of the South Georgia is covered by permanent ice with many large glaciers reaching tidewater. The island is oriented NW to SE and is 170 km long and varies in width from 2 to 40 km. The British Antarctic Survey (BAS) has been the principal research group examining glacier change on South Georgia Island. Cook *et al.* (2010) and Gordon, Haynes, and Hubbard (2008) have emphasized that there is an island-wide pattern with many calving glaciers having faster retreat. Gordon, Haynes, and Hubbard (2008) observed that larger tidewater and sea-calving valley and outlet glaciers generally remained in relatively advanced positions from the 1950s until the 1980s. After 1980, most glaciers receded; some of these retreats have been dramatic and a number of small mountain glaciers will soon disappear. Cook *et al.* (2010) completed a series of BAS maps that map the changes in glacier front from 1958 to 2007. This study notes that 97% of the 102 coastal glacier retreated between 1950 and 2007. Cook *et al.* (2010) identified that 212 of the Peninsula's 244 marine glaciers have retreated over the past 50 years and that rates of retreat are increasing. In the, who would have ever guessed it category, the glacier retreat has been an aid to the rat population, as the glacier tongues used to corner populations (Cook *et al.*, 2010) (Fig. 5.1).

Heard Island is at 53 S and 72 E and is dominated by a large volcano, Big Ben. The Australian Antarctic Division manages Heard Island and has undertaken a project documenting the changes in the environment on the island. One aspect noted has been the change in glaciers. The Allison, Brown, and Stephenson Glaciers have all retreated substantially since 1947 when the first good maps of their terminus are available. *Fourteen Men* by Arthur Scholes (1952) documents a year spent by 14 men of the Australian National Antarctic Research Expedition that documented the particularly stormy, inclement weather of the region. Their visit to the glacier noted that they could not skirt past the Stephenson Glacier along the coast. After crossing Stephenson Glacier, they visited an old seal camp and counted 16,000 seals in the area. Thost and Truffer (2008) noted a 29% reduction in area of the Brown Glacier from 1947 to 2003. They also observed that the volcano Big Ben that the glaciers all drain from has shown no sign of changing geothermal output to cause the melting and that a 1 °C

Recent Climate Change Impacts on Mountain Glaciers, First Edition. Mauri Pelto.
© 2017 John Wiley & Sons, Ltd. Published 2017 by John Wiley & Sons, Ltd.

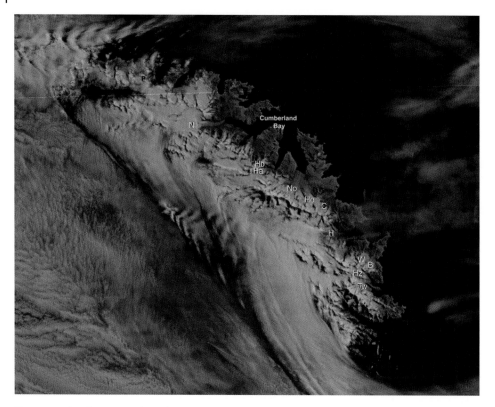

Figure 5.1 South Georgia glacier locations on 2015 satellite image. Tw = Twitcher, Hz = Herz, B = Bertrab, W = Weddell, H = Hindle, C = Cook, He = Heaney, No = Nordenskjold, Ha = Harker, Hb = Hamberg, N = Neumayer, K = Konig, and P = Purvis.

warming has occurred over the same time period Kiernan and McConnell (2002). Sea surface temperatures in the region have also risen, and given the limited tidewater nature of the glaciers on Heard Island, this would have a limited role in overall retreat except for at Stephenson Glacier with its large lagoon.

Kerguelen Island is in the southern Indian Ocean (49°S, 69°E), and the principal area of glaciers is the Cook Ice Cap. A comparison of aerial images from 1963 and 2001 by Berthier *et al.* (2009) indicated that the Cook Ice Cap had lost 21% of its area in the 38-year period. A ~1 °C warming has been noted for the region since 1964 (Berthier *et al.*, 2009). The glaciers are not tidewater, and a warmer ocean will not impact the glaciers.

5.1 Twitcher Glacier

Twitcher Glacier is the next glacier south of Herz Glacier on the east coast of South Georgia (Fig. 5.2). Until 1989 the glacier ended at the tip of a peninsula, and the ensuing retreat has led to the opening of a new fjord. Twitcher Glacier was 10 km long and had a 2 km wide calving front in 2009. The terminus change of this tidewater glacier was completed by the BAS for the 1960–2007 period, and the glacier had retreated 1.5 km, with most of the retreat occurring after 1992 (Gordon, Haynes, and Hubbard, 2008).

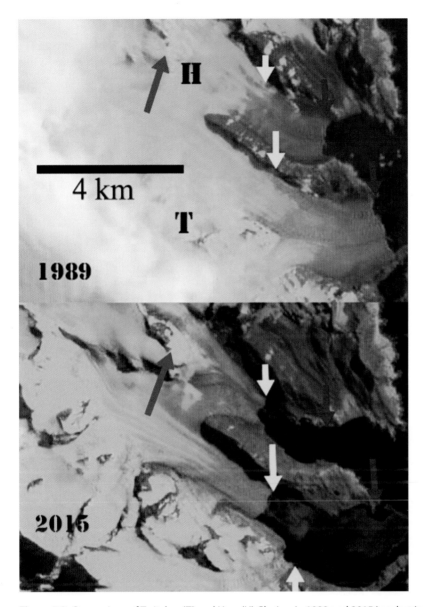

Figure 5.2 Comparison of Twitcher (T) and Herz (H) Glaciers in 1989 and 2015 Landsat images. Red arrows indicate 1989 terminus locations, yellow arrows 2015 terminus locations, and purple arrows upglacier thinning.

In 1989 this glacier terminated approximately at the end of a peninsula separating the two glaciers. Here we examine Landsat imageries from 1989, 2000, 2009, and 2015 to identify the rate retreat. The 1989 and the 2009 terminus positions are indicated a retreat of 1.2 km during this period. Respectively. The 1989 and the 2009 terminus positions are indicated a retreat of 1.2 km during this period. From 2009 to 2015, retreat accelerated with a further 1.7 km retreat to the red arrow in the 2015 imagery. The terminus is currently quickly retreating to the next peninsula where the terminus will separate into two parts. The southern tributary already is partly exposed to calving into the fjord.

5.2 Herz Glacier

Herz Glacier is on the southeast coast of South Georgia Island (Fig. 5.2). The terminus change of this tidewater glacier ending in Iris Bay was completed for the 1960–2007 period, with a slow retreat from 1960 to 1988 and a more rapid retreat since (Gordon, Haynes, and Hubbard, 2008).

Here we examine imagery from Google Earth and Landsat to examine terminus change from 1989 to 2015. The terminus on the north side of the fjord has retreated 1.8 km in the 25-year period, and the terminus on the south side has retreated 2.2 km. The overall 2 km retreat is a rate of 80 m year^{-1} and is 20% of the total glacier length.

5.3 Weddel Glacier

Weddel Glacier is on the southeast coast of South Georgia Island (Fig. 5.3). It terminates in Beaufoy Cove just north of Gold Harbor. In 1958 it reached within 400 m of the coast at the outlet of Beaufoy Cove. For Weddel Glacier, the retreat was rapid from 1960 to 1974 and was slow from 1992 to 2003 (Cook *et al.*, 2010). Here we examine Landsat imageries from 1989 to 2015 to visualize and update this change.

In 1989 the glacier terminates near the tip of a peninsula (red arrow in each image; Fig. 5.4). The calving front extends southeast. In 2002 there is only minor retreat, but thinning has led to the small extension of the main icefall being almost cut off by bedrock. In 2015 the glacier has retreated 200–300 m from the 1989 position, and the main terminus is narrower. A Google Earth close-up of

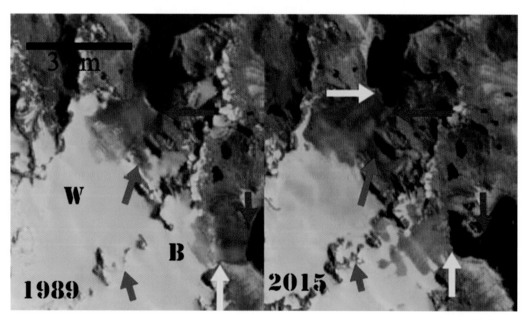

Figure 5.3 Comparison of Weddel (W) and Bertrab (B) Glaciers in 1989 and 2015 Landsat images. Red arrows indicate 1989 terminus locations, yellow arrows 2015 terminus locations, and purple arrows upglacier thinning.

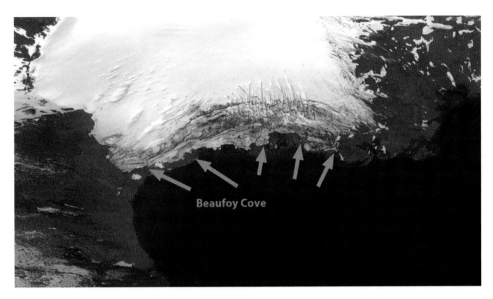

Figure 5.4 Google Earth 2009 image. (Google Earth.)

the terminus indicates that only a small section is still in contact with Beaufoy Cove in 2009, with land exposed at the orange arrows (Fig. 5.4). This glacier is almost not tidewater and has terminated in shallow water since 1989, which helps explain a slower rate of retreat. The glacier has thinned more rapidly than it has retreated in the last 25 years.

5.4 Bertrab Glacier

Bertrab Glacier is on the east coast of South Georgia Island. In 1958 it reached the coast in Gold Harbor. For Bertrab Glacier, the retreat was minimal from 1958 to 1989. Since 1989, a whole new embayment has opened (Cook *et al.*, 2010). Here we examine Landsat imageries from 1989 to 2015 to visualize and update this change.

In 1989 the southern arm of the glacier extends to the shoreline of the barrier beach system in Gold Harbor (red arrow; Fig. 5.3). The northern arm extends around to the edge of a very green region, suggesting well-developed vegetation, hence no real retreat. In 2002 a lake has formed at the northern arm terminus, and it has retreated 400 m. The southern arm has retreated across a new embayment, though the exact position is obscured by cloud. In 2011 the southern terminus has retreated up a slope from the edge of the embayment. In 2015 there are no longer two arms to the glacier. The glacier terminates near the edge of the new embayment. The retreat is 700 m on the northern arm and 1000 m for the southern end since 1989. The glacier no longer reaches the water limiting calving. The glacier also ends on moderate slope. This should lead to a reduced retreat in the near future. The 2015 image is from January 15, so there is still 2 months left in the melt season.

Like on Stephenson Glacier, Heard Island, the new embayment does offer new potential habitat for penguins and seals.

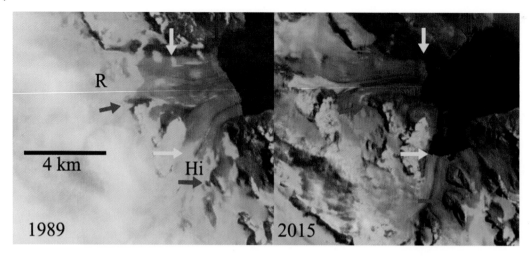

Figure 5.5 Comparison of Ross (R) and Hindle (Hi) Glaciers in 1989 and 2015 Landsat images. Red arrows indicate 1989 terminus locations, yellow arrows 2015 terminus locations, and purple arrows upglacier thinning.

5.5 Ross–Hindle Glacier

The Ross–Hindle Glacier that enters Royal Bay on the east coast of South Georgia Island has now separated into the Ross and Hindle Glaciers. In 1989 the glaciers are joined and remain so despite a 1 km retreat by 2003 (Fig. 5.5).

In 2008 the glaciers had separated. Gordon, Haynes, and Hubbard (2008) observed that larger tidewater and sea-calving valley and outlet glaciers generally remained in relatively advanced positions until the 1980s. For Ross–Hindle Glacier, the retreat was minimal from 1960 to 1989. Hindle Glacier is on the verge of separating again into a west and east arm. There is an increase in surface slope within 1 km of the current terminus, indicating retreat should slow after that. Ross Glacier is retreating into a widening valley, which provides no evident pinning point to slow retreat. There is no large increase in surface slope indicating a decrease in water depth either.

From 2003 to 2015 retreat has accelerated with the Hindle Glacier retreating south into a separate fjord. In 2015 Hindle Glacier has retreated 2.75 km and Ross Glacier 1.3 km since 2003. This is 100 m year^{-1} for Ross and 200 m year^{-1} for Hindle Glacier. The exceptional retreat rate of Hindle Glacier suggests that Ross Glacier acted as a pinning point stabilizing the terminus reach of the glacier.

5.6 Heaney Glacier–Cook Glacier

Heaney Glacier and Cook Glacier were merged near the terminus and reached the coast of St. Andrews Bay (Fig. 5.6) on the east coast of South Georgia Island in 1975. This is illustrated in a Geomorphology map of the area compiled by Clapperton and Sugden produced with the support of the BAS; the glaciers terminate along the coastline.

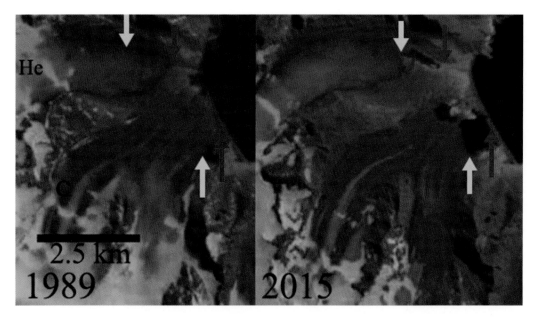

Figure 5.6 Comparison of Cook (C) and Heaney (He) Glaciers in 1989 and 2015 Landsat images. Red arrows indicate 1989 terminus locations, yellow arrows 2015 terminus locations, and purple arrows upglacier thinning.

Here we use Landsat images to examine glacier change from 1989 to 2015. In 1989, Cook Glacier had retreated from the coast, and a small 200 – 300 m wide proglacial lake has formed at the end of the glacier (red arrow). The red arrow marks the 1989 terminus of Heaney Glacier, which is 800 m from the coast. The yellow arrow indicates 2015 terminus position. In 1999 Landsat imagery indicates modest retreat of both glaciers. In 2003 Cook Glacier has retreated 500 m from the coastline, and Heaney Glacier is now 1100 m from the coast. In 2012 a small lake is developing at the front of the Heaney Glacier, and the Cook Glacier proglacial lake has expanded to 700 m. In 2015 the narrow lake forming as Heaney Glacier's retreat is now 600 m long, and the glacier terminus is 1800 m from the coast (pink arrow). This is a 1000 m retreat from 1989 to 2015, 40 m a^{-1}. Point J is now fully deglaciated with Cook and Heaney Glacier being fully separated. Cook Glacier has retreated 900 m from the coast and 600 – 700 m since 1989. The proglacial lake (red arrow) is 750 m across and is still expanding as the glacier retreats.

5.7 Nordenskjold Glacier

Nordenskjold Glacier has a 3 km wide calving front in Cumberland East Bay. The glacier experienced minimal retreat from 1957 to 2003 (Gordon, Haynes, and Hubbard, 2008). From 1989 to 2015, the glacier has retreated 1000 m, which is less than most major tidewater outlet glaciers of South Georgia (Fig. 5.7). The fjord widens significantly 1 km behind the current front. Once the glacier retreats beyond this point, retreat will become rapid as the widening fjord provides a reduced pinning point. There is upglacier thinning at the purple arrows indicating that the retreat will be ongoing.

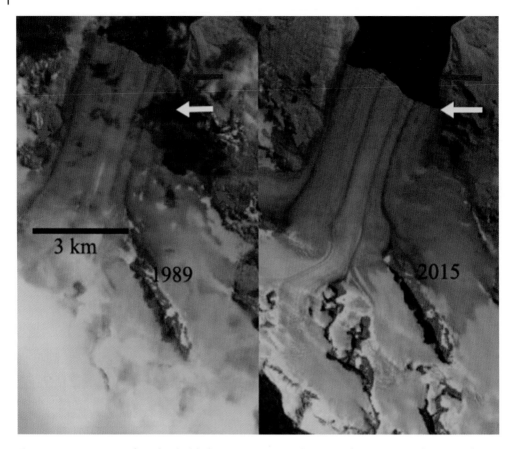

Figure 5.7 Comparison of Nordenskjold Glacier (C) in 1989 and 2015 Landsat images. Red arrows indicate 1989 terminus locations, yellow arrows 2015 terminus locations, and purple arrows upglacier thinning.

5.8 Harker and Hamberg Glaciers

Harker and Hamberg Glaciers are tidewater glaciers are at the head of Moraine Fjord. The glaciers were at the mouths of their respective fjords in 1957 (Clapperton, Sugden, and Pelto, 1989). In 1989 Hamberg Glacier had retreated a short distance west into an arm of the fjord (Fig. 5.8). Harker Glacier remained pinned at the entrance to the southern arm of the fjord. Harker Glacier has retreated 1400 m as the southern arm of the fjord has opened. The glacier is retreating into a narrower fjord reach, which should slow retreat in the short run. In 2015 Hamberg Glacier had retreated 900 m from the 1989 position and remains a tidewater glacier. The width and depth of the fjord have both declined, slowing the calving-driven retreat of Hamberg Glacier.

5.9 Neumayer Glacier

Clapperton, Sugden, and Pelto (1989) noted the equilibrium line altitude (ELA) of Neumayer Glacier at 550 m. Neumayer Glacier is one of the large tidewater glaciers on South Georgia. Maps

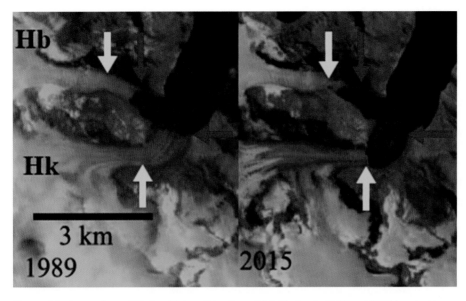

Figure 5.8 Comparison of Harker (Hk) and Hamberg (Hb) Glaciers in 1989 and 2015 Landsat images. Red arrow indicates 1989 terminus locations and yellow arrows 2015 terminus.

from the BAS and satellite images are used here to assess the changes in this glacier terminus position.

The BAS has a mapping function that provides glacier front positions since early in the twentieth century. For Neumayer Glacier, the 1938 position is 3.5 km down fjord from the 2006 position. There was essentially no retreat up to 1974 and limited retreat up to 1993 (Fig. 5.9) (Gordon *et al.*, 2008). Landsat images from 1999 to 2015 indicate retreat of 4800 m from the red to the yellow arrow; this is 320 m year^{-1} (Fig. 5.9). The glacier appears to have retreated into a deeper section of the fjord then where it ended from 1970 to 2002. This will enhance calving from the glacier and promote additional mass loss and retreat. This retreat will impact Konig Glacier, which is connected to the Neumayer Glacier. Calving rate increases with water depth and the degree of glacier buoyancy (Brown *et al.*, 1982; Pelto and Warren, 1991).

5.10 Konig Glacier

Konig Glacier is a land-terminating glacier just north of the Neumayer Glacier, ending on an outwash plain in the Antarctic Bay (AB) on the northwest coast of South Georgia. In 1977 the glacier extended to within 300 m of Antarctic Bay, and no proglacial lake existed. Neumayer Glacier is a calving glacier that has retreated 4800 m from 1999 to 2015 and is dynamically connected to the Konig Glacier along its southern margin just where the glacier turns northeast.

Here we examine the changes in Konig Glacier from 1999 to 2014 using Landsat imagery (Fig. 5.10). In 1999 the glacier ended in a proglacial lake at the red arrow, where a terminal moraine developed across the lake. A tributary glacier from the west joins the Konig Glacier near the terminus in 1999 (green arrow). At the right-hand side, the purple arrow indicates a small cirque-valley glacier that

Figure 5.9 Comparison of Neumayer Glacier in 1989 and 2015 Landsat images. Red arrows indicate 1989 terminus locations, yellow arrows 2015 terminus locations, and purple arrows upglacier thinning.

joins the Neumayer Glacier near the boundary with Konig Glacier. In 2003 there has been limited retreat of the main terminus since 1999 and of the west tributary. The side cirque glacier at the green arrow is still connected. In 2005 a close-up of the terminus in Google Earth indicates the low slope, lack of crevasses, and developing outwash plain at the terminus. The terminal moraine in the middle of the lake marking the 1993 terminus position is also evident. In 2014 the glacier has retreated to the yellow arrow, which is an 800 m retreat in 15 years from the 1999 red arrow terminus. The proglacial

Figure 5.10 Comparison of Konig Glacier in 1989 and 2015 Landsat images. Red arrows indicate 1989 terminus locations, yellow arrows 2015 terminus locations, and purple arrows upglacier thinning.

lake is now 1500 m across, and the terminus is 2300 m from Antarctic Bay. The west tributary at the purple arrow is fully separated.

5.11 Purvis Glacier

Purvis Glacier is on the northeastern coast of the island, terminating on land near Possession Bay. The Cook *et al.* (2010) mapping indicated in Fig. 5.11 shows that the Purvis Glacier terminus was on the coastline in 1974. Here we examine Landsat images from 1999 to 2014 to identify more recent changes.

In 1999 the proglacial lake (red arrow) that the glacier terminated in was 300 m wide, indicating a retreat of 300–400 m since 1974 (Fig. 5.11). In 2002 the proglacial lake had expanded to a width of 600 m. In March 1, 2014, Landsat imagery indicates a retreat of 1100 m since 1974, with most of that retreat occurring since 1989 (Fig. 5.11). A closer look at the glacier from Google Earth highlights the issue (Fig. 5.12). The glacier is fed by relatively low-lying snowfields with quite limited areas above 500 m. Clapperton, Sugden, and Pelto (1989) identified the snow line a short distance from here at 400–450 m. As the 2011 Google Earth image indicates, the remaining snow cover at the end of the melt season is minimal, too little to sustain this glacier (Pelto, 2010). Further a look at the terminus indicates the stagnant nature of the terminus region that will lead to continued retreat; blue arrows note ablation holes in the glacier that do not develop when a glacier is actively moving. The low slope and stagnant nature should preserve an excellent glacial geologic landscape.

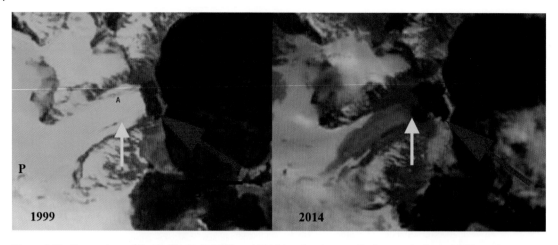

Figure 5.11 Comparison of Purvis Glacier in 1989 and 2015 Landsat images. Red arrows indicate 1999 terminus locations, yellow arrows 2014 terminus locations, and Point A represents a peninsula in the developing lake.

Figure 5.12 Google Earth image in 2011. (Google Earth.)

5.12 Stephenson Glacier–Heard Island

Stephenson Glacier extends 8–9 km down the eastern side of Big Ben (Fig. 5.13). In 1947 it spread out into a piedmont lobe that was 3 km wide and extended to the ocean in two separate lobes around Elephant Spit.

Kiernan and McConnell (2002) identified an order of magnitude increase in the rate of ice loss from Stephenson Glacier after 1987. Retreat from the late nineteenth century to 1955 had been limited.

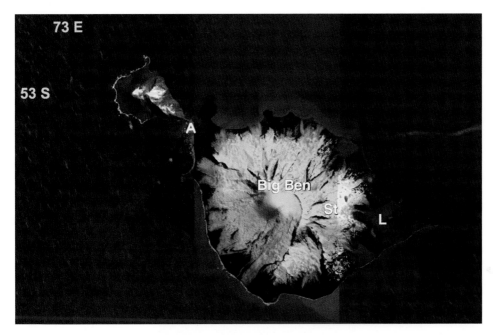

Figure 5.13 Heard Island overview from Google Earth images. A = Atlas Cove, St = Stephenson Glacier, and L = Stephenson Lagoon. (Google Earth.)

As Kiernan and McConnell (2002) observed, retreat began by 1971 the glacier had retreated 1 km from the south coast and several hundred meters from the northern side of the spit. This retreat in 1980 caused the formation of Stephenson Lagoon, and by 1987, Doppler Lagoon had formed as well. After 2000, the two lagoons have joined.

Here we compare Landsat images from 2001 and 2013 to update the response of this glacier (Fig. 5.14). In 2001 the glacier has two separate termini (purple arrows) in two different lagoons: Doppler to the south and Stephenson to the east. There are numerous icebergs in Doppler Lagoon but none in Stephenson Lagoon, indicating the retreat is underway. In 2008 the two lagoons are well joined; icebergs are even more numerous, obscuring in this view the true location of the terminus. In 2010 the glacier has retreated from the main basin of the lagoon, and the lagoon is free of ice for the first time in several centuries if not several millennium. In 2013 the glacier has retreated into a narrower valley that feeds into Stephenson Lagoon. The northern arm of the glacier experienced a 1.7 km retreat from 2001 to 2013 and the southern arm 3.4 km retreat. The period of rapid retreat due to calving of icebergs into the lagoon is over and the retreat rate will now be slower. There is still rapid glacier flow toward the terminus as indicated by extensive crevassing. The overall glacier slope is steep and accumulation rates high, which would also generate rapid glacier flow. In the 2014 Google Earth image, the lack of icebergs in the lagoon indicates another transition in this new ecosystem (Fig. 5.15).

The climate station at Atlas Cove indicates a 1 °C temperature rise in the last 60 years (AAD, 2015). The AAD (2015) will certainly be looking at how this new lagoon impacts the local seal and penguin communities.

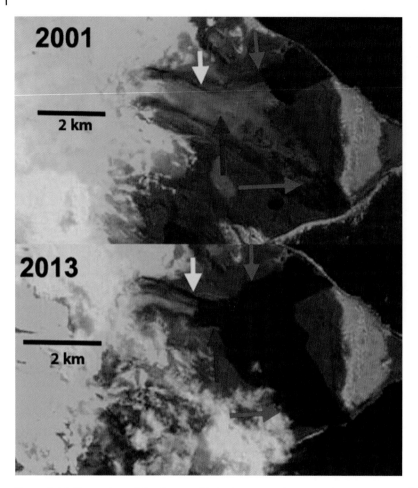

Figure 5.14 Comparison of Stephenson Glacier in 2001 and 2013 Landsat images. Purple arrows indicate 2001 terminus, red arrows 2007 terminus, and yellow arrows 2013 terminus location.

5.13 Agassiz Glacier–Kerguelen Island

Kerguelen Island sits alone at the edge of the furious fifties in the southern Indian Ocean. The island features numerous glaciers, the largest being the Cook Ice Cap at 400 km^2 (Fig. 5.16).

A comparison of aerial images from 1963 and 2001 by Berthier *et al.* (2009) indicated that the ice cap had lost 21% of its area in the 38-year period. Berthier *et al.* (2009) in a related part of the aforementioned study noted that the retreat accelerated after 1970 and again after 2003, with thinning of 5 m a^{-1} on Ampere Glacier and retreat of 75 m a^{-1} on Explorer Glacier on the east side of the ice cap.

A comparison of Google Earth (top), 2001 (middle), and 2011 (bottom) Landsat images indicate a significant retreat and formation of a new lake that is not evident in maps or Google Earth imagery. The red arrows indicate the 2001 terminus position with no proglacial lake (Fig. 5.17). In 2011 a lake has formed and the terminus has retreated to the yellow arrow. Agassiz Glacier has lost its field hockey

Figure 5.15 Stephenson Glacier (right) and Stephenson Lagoon in 2014 Google Earth image. The lagoon is nearly iceberg free. (Google Earth.)

stick–shaped hook and has retreated 2 km in just 10 years; the rate of 200 m a^{-1} is quite high even by Cook Ice Cap standards. The glacier is calving into the lake that can accelerate retreat. The terminus does not appear to be at a stable point, with a wide calving front in comparatively deep water.

5.14 Ampere Glacier

Ampere Glacier is the most prominent outlet glacier of the Cook Ice Cap (Fig. 5.18). Berthier *et al.* (2009) noted a retreat from 1963 to 2006 of 2800 m of the main glacier termini in Ampere Lake (As). The lake did not exist in 1963. A second focus of their work was on the Lapparent Nunatak due north of the main terminus and close to the Ampere Glacier's east terminus (Ae). The nunatak expanded from 1963 to 2001, in the middle image from Berthier *et al.* (2009), but it was still surrounded by ice.

The main terminus has retreated 800 m from 2001 to 2013. Here the terminus has pulled back from the tip of the peninsula on the west side of the terminus and is currently at a narrow point. The eastern terminus has retreated to its junction, with the main Ampere Glacier to a distance of 1400 m. Berthier *et al.* (2009) had noted thinning around the Lapparent Nunatak of 150–250 m; purple arrows indicate this location of thinning. Above the current main terminus, the valley widens again to the junction with the location of the eastern terminus. It seems likely that the main glacier will retreat north until there is a single terminus north of the southern end of Lapparent Nunatak.

5.15 Lapparent Glacier

Lapparent Glacier is a major outlet glacier on the south side of the Cook Ice Cap. Landsat imageries from 2001 to 2013 indicate widespread thinning and deglaciation of this glacier (Fig. 5.18). In 2001

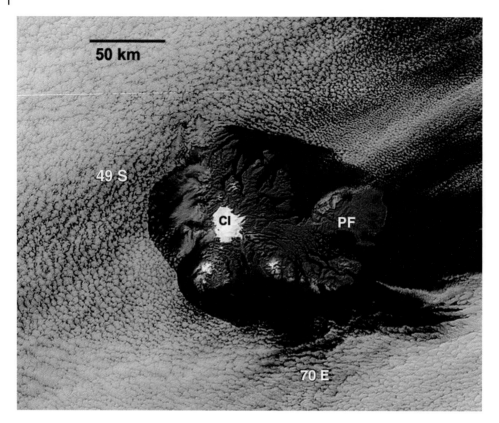

Figure 5.16 Kerguelen Island in 2011 MODIS image. CI = Cook Ice Cap and PF = Port du Francais. (Jeff Schmaltz, MODIS Land Rapid Response Team, NASA GSFC.)

Lapparent Glacier merges with the east terminus of Ampere Glacier at the red arrows with a medial moraine evident. In 2013 the eastern arm has narrowed from 1100 to 500 m and retreated 2100 m in 12 years. The result is less ice flow over a bedrock step just above the terminus. This continued thinning since 2001 will lead to further retreat of the glacier. There is no calving and the rate of retreat will decline.

5.16 Lake District

On the east side of the Cook Ice Cap on Kerguelen Island, a series of outlet glaciers have retreated expanding and forming a new group of lakes. This area is just south of Lac du Chamonix, which has existed. Here we examine the changes from 2001 to 2014 using Landsat imagery (Fig. 5.19).

In 2001 the southern outlet glacier terminates at the northeastern end of the lake. In 2014 the glacier has retreated 900 m to the yellow arrow, with a new lake developing. In 2001 the next glacier north terminates at the eastern end of a lake basin, at the red arrow. In 2014 the glacier has retreated 600 m opening up a new lake. In the midst of the third lake is an island, marked with point A. From 2001 to

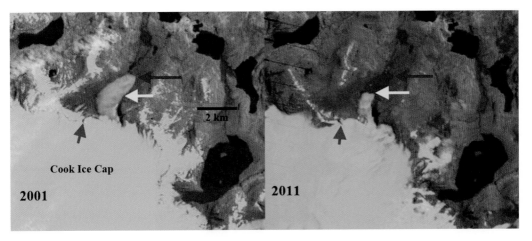

Figure 5.17 Comparison of Agassiz Glacier, northern outlet of Cook Ice Cap in 2001 and 2011 Landsat images. Red arrows indicate 2001 terminus locations, yellow arrows 2011 terminus locations, and purple arrows upglacier thinning.

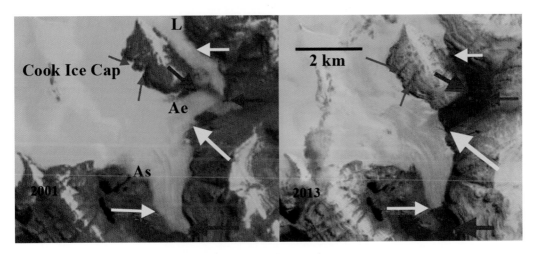

Figure 5.18 Comparison of Ampere (A) and Lapparent (L) Glaciers, southern outlet glaciers of the Cook Ice Cap in 2001 and 2013 Landsat images. Red arrows indicate 2001 terminus locations, yellow arrows 2013 terminus locations, and purple arrows upstream thinning.

2014, the glacier has retreated 400 m from this island. In 2001 the northern tributary terminates at the red arrow. In 2014 the glacier has retreated into a narrow inlet, indicating the lake basin is nearing its western end. The glacier has separated from a former tributary at the purple arrow.

In just a decade we see the formation of two new lakes and the expansion of two others at the terminus of the three eastern outlet glaciers of Cook Ice Cap.

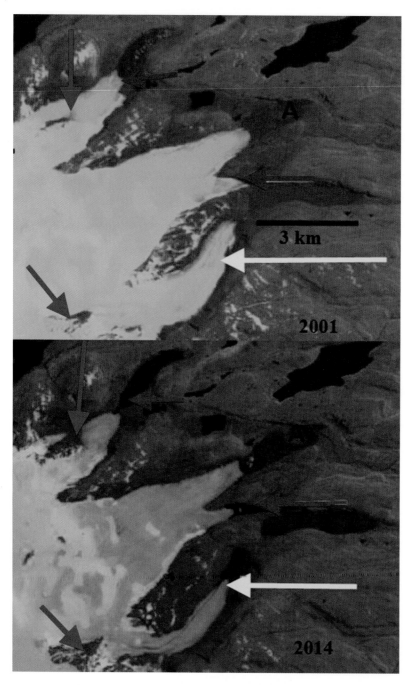

Figure 5.19 Comparison of eastern outlet glaciers of the Cook Ice Cap in 2001 and 2014 Landsat images. Red arrows indicate 2001 terminus locations, yellow arrows 2013 terminus locations, and purple arrows upstream thinning.

References

Australian Antarctic Division (2015) *Heard Island and McDonald Island Marine Reserve Research*, http://heardisland.antarctica.gov.au/research (accessed December 2015).

Berthier, E., Lebris, R., Mabileau, L. *et al.* (2009) Ice wastage on the Kerguelen Islands (49S, 69E) between 1963 and 2006. *JGR-Earth Surface*, **114**, F03005. doi: 10.1029/2008JF001192

British Antarctic Survey (2010) *South Georgia Geographic Information System (GIS)*, http://geo.antarctica.ac.uk/ (accessed June 2015).

Brown, C., Meier, M., and Post, A. (1982) Calving speed of Alaska tidewater glaciers, with application to Columbia Glacier. USGS Prof. Pap.1258-C.

Clapperton, C.M., Sugden, D.E., and Pelto, M. (1989) Relationship of land terminating and fjord glaciers to Holocene climatic change, South Georgia, Antarctica, in *Glacier Fluctuations and Climatic Change* (ed. J. Oerlemans), Kluwer, Dordrecht, pp. 57–75.

Cook, A., Poncet, S., Cooper, A. *et al.* (2010) Glacier retreat on South Georgia and implications for the spread of rats. *Antarctic Science*, **22**, 255–263. doi: 10.1017/S0954102010000064

Gordon, J.E., Haynes, V., and Hubbard, A. (2008) Recent glacier changes and climate trends on South Georgia. *Global and Planetary Change*, **60** (1–2), 72–84.

Kiernan, K. and McConnell, A. (2002) Glacier retreat and melt-lake expansion at Stephenson Glacier. *Heard Island World Heritage Area Polar Record*, **38** (207), 297–308.

Pelto, M.S. (2010) Forecasting temperate alpine glacier survival from accumulation zone observations. *The Cryosphere*, **3**, 323–350.

Pelto, M. and Warren, C. (1991) Relationship between tidewater glacier calving velocity and water depth at the calving front. *Annals of Glaciology*, **15**, 115–118.

Scholes, A. (1952) *Fourteen Men*, E. P. Dutton, New York, NY.

Thost, D.E. and Truffer, M. (2008) Glacier recession on Heard Island, Southern Indian Ocean, Arctic. *Antarctic and Alpine Research*, **40** (1), 199–214.

6

Svalbard: Hornsund Fjord Region

Overview

Svalbard is host to 163 tidewater glaciers with a collective calving front of 860 km (Blaszczyk, Jania, and Hagen, 2009). They observed that 14 glaciers had retreated from the ocean to the land over the last 3–4 decades. Nuth *et al.* (2013) determined that the glacier area over the entire archipelago has decreased by an average of 80 km^2 a^{-1} over the past 30 years, a 7% reduction. In the most recent period, 1990–2007, terminus retreat was larger than in an earlier period from 1930 to 1990, while area shrinkage was smaller.

Most of the large glaciers in Svalbard are tidewater calving glaciers. Svalbard glaciers have been losing considerable volume, indicative of negative mass balance and glacier retreat. Nuth *et al.* (2010) concluded that over the past 40 years for Svalbard ice loss is 9.71 ± 0.55 km^3 a^{-1}. This is an average thinning of 0.36 m a^{-1} for an annual contribution to global sea-level rise of 0.026 mm a^{-1}.

In 2011 sea surface temperatures were 5 °C above the mean in the Barents Sea; this has been a sustained warming in recent years (Walczowski and Piechura, 2011). Sea ice extent has both been reduced by ocean warming and enhanced the warming. A sequence of sea ice images indicates the change in sea ice extent in the region during the 2004–2012 period (Fig. 6.1). Warmer ocean water is a key control of tidewater calving rates (Luckman *et al.*, 2015).

In this chapter, we focus on the glaciers of Hornsund Fjord that in 2014 almost cuts through the southern island of Svalbard (Fig. 6.2). The Landsat images are from August 20, 1990 (LT5280051990232KIS00) and August 4, 2014 (LC82100052014216LGN00). The Institute of Geophysics Polish Academy has maintained a Polish Research Station in Hornsund since 1957. The 1984 map, from the University of Silesia, of the glaciers and geomorphology document the extent of the glaciers in 1983. A recent detailed examination by Blaszczyk, Jania, and Kolondra (2013) reported the total area of the glacier cover lost in Hornsund Fjord area from 1899 to 2010 was approximately 172 km^2. The average glacier area retreat increased from a mean of 1.6 to 3 km^2 a^{-1} since 2000. Water temperature in Hornsund combined with the sea ice in the fjord that is also crucial in controlling the development of waterline notches that facilitates calving (Petlicki *et al.*, 2015).

6.1   South Coast of Hornsund

There are five glaciers along the south coast of Hornsund Fjord. From east to west, they are Svalisbreen, Mendeljevbreen, Chomjakovbreen, Samarinbreen, and Korberbreen (Fig. 6.3). In the

Recent Climate Change Impacts on Mountain Glaciers, First Edition. Mauri Pelto.
© 2017 John Wiley & Sons, Ltd. Published 2017 by John Wiley & Sons, Ltd.

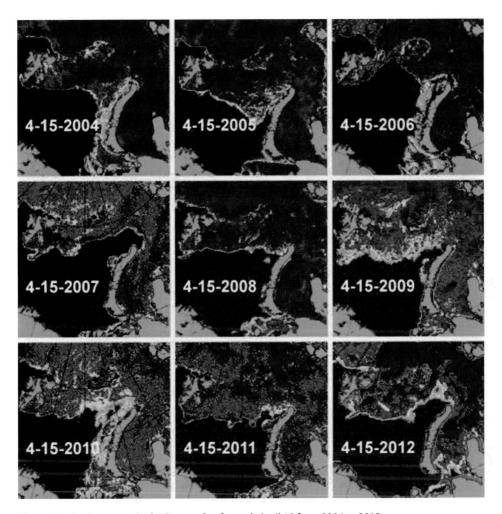

Figure 6.1 Sea ice extent in the Barents Sea for each April 15 from 2004 to 2012.

1990 Landsat image, the terminus is denoted by a red arrow and red dots. In the 2014 Landsat image, the terminus is indicated by yellow dots, with the 1990 terminus position indicated by red arrows. The most significant retreat is 2400 m for Svalisbreen, which in 1990 reached the tip of the peninsula both on the north with Hambergbreen and south with Mendeljevbreen, but by 2014 it had retreated east up the new arm of fjord that appears will widen further with additional retreat. Mendeljevbreen had been joined with Svalisbreen in 1990 and had by 2014 retreated 1300 m up its own fjord. Chomjakovbreen has retreated 500 m with the greatest retreat in the glacier center. Chomjakovbreen has the narrowest active calving front. The east and west margins of the glacier remain pinned on the fjord walls, indicating the narrow fjord reach. The retreat of Samarinbreen has been 1200 m and with the glacier retreating from a pinning point at the 1990 terminus area, the glacier is poised to continue rapid retreat in the wider fjord. Korberbreen has retreated 600 m. All five remain calving glaciers and will continue to experience a retreat enhanced by calving.

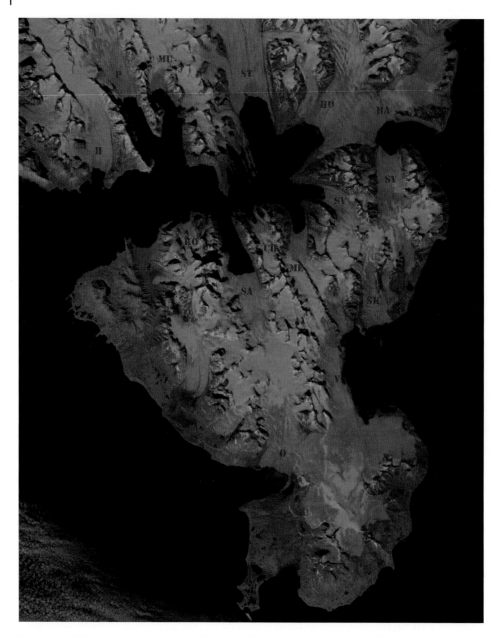

Figure 6.2 A 2014 Landsat image in the Hornsund Fjord region of southern Svalbard in 1990 provides a useful snapshot of glacier change in the region. The glaciers examined here are H = Hornsund, P = Paierbreen, MU = Muhlbacherbreen, ST = Storbreen, HO = Hornbreen, HA = Hambergbreen, SY=, SV = Svalisbreen, ME = Mendeljevbreen, CH = Chomjakovbreen, SA = Samarinbreen, KO = Korberbreen, SK = Skilfonna, V = Vasilievbreen, and O = Olsokbreen.

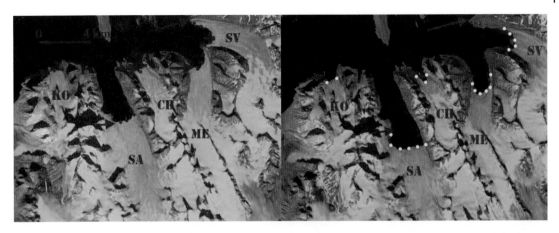

Figure 6.3 Landsat image analysis of 1990 and 2014 of KO = Korberbreen, SA = Samarinbreen, ME = Mendeljevbreen, CH = Chomjakovbreen, and SV = Svalisbreen.

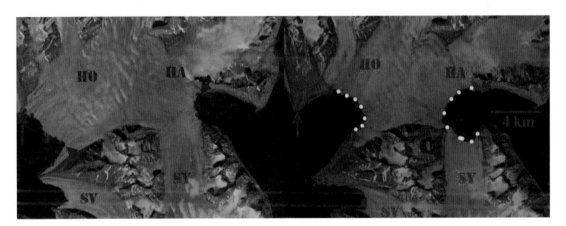

Figure 6.4 Eastern end of Hornsund with Hornbreen (HO), Svalisbreen (SV), Sykorabreen (SY), and Hambergbreen (HA).

6.2 Eastern Hornsund Glacier change

In 1984, the eastern end of Hornsund was fronted by a single glacier terminus comprised of the Storbreen (ST), Hornbreen (HO), Svalisbreen (SV), and Mendeljevbreen (Fig. 6.4). By 1990, the glaciers had all separated.

Palli *et al.* (2003) in a joint University of Oulu, Finland and University of Silesia noted that Hornbreen has retreated by 13.5 km from the central part of the front and Hambergbreen by 16 km from 1901 to 2000. As part of this project surveyed the basal topography beneath the glaciers (Palli *et al.*, 2003), they found that there is not a below sea-level connection underneath the Hornbreen–Hambergbreen divide that would separate Sorkappland from Torrelland. The ice divide of Hornbreen–Hambergbreen is below the local snow line at 300 m, and Palli *et al.* (2003) indicate that this connection cannot survive current climate. In both 1990 and 2014, the divide is well below the snow line. Kvamstø *et al.* (2012) in a Bergen University–led study noted that the melt season

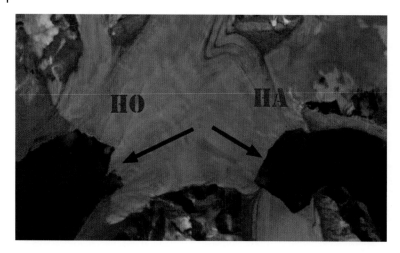

Figure 6.5 2010 Landsat image of area from the front of Hornbreen to Hambergbreen.

Figure 6.6 The north side of Hornsund with from west to east Hansbreen (H), Paierbreen (P), Muhlbacherbreen (MU), and Storbreen (ST).

had increased by more than 2 weeks in Svalbard from 1912 to 2010. In 1983, the distance from the terminus of the Hornbreen to the terminus of Hambergbreen was 17 km. In 2013, the distance is 9 km (Fig. 6.5). The divide is below the snow line in both 2013 and 2014, not a sustainable dynamic setup. A comparison of locations in the 1990 and 2014 Landsat images indicates that the retreat has been approximately 3.2 km for Hornbreen and 2.7 km for Hambergbreen. Hambergbreen has also separated from Sykorabreen, which has retreated 800 m. This has expanded the calving front, which will lead to more ice loss.

6.3 North side of Hornsund

From 1990 to 2014, all four of the glaciers Hansbreen (H), Paierbreen (P), Muhlbacherbreen (MU), and Storbreen (ST), ending on the north coast of Hornsund, have retreated significantly (Fig. 6.6). Storbreen has retreated the most with 3600 m of retreat, and the fjord arm that has been extended by the retreat is maintaining its width indicating a lack of pinning points to slow the retreat. Muhlbacherbreen has retreated 2100 m, retreating from a pinning point in 1990 where the fjord was narrower. The glacier slope is much steeper than on Storbreen, and the fjord arm may not extend much further. Paierbreen has retreated 2200 m with the current terminus in a narrow point in the fjord. Beyond this location, the fjord expands again, which would enhance calving and retreat. Hansbreen has been examined in detail over the last 20 years from the Polish Research Station. Vieli *et al.* (2002) noted that retreat rate varied more due to buoyancy changes due to depth changes then from mass balance. The glacier retreated 2700 m from 1900 to 2008 (Oerlemans, Jania, and Kolondra, 2011). They illustrate this retreat as does the comparative images from the Polish Research Station. The glacier mass balance has been measured since 1989 and is submitted to the World Glacier Monitoring Service. In a detailed review of this calving glacier, Oerlemans, Jania, and Kolondra (2011) report that the average surface mass balance has been $-0.36\,\mathrm{m\,a^{-1}}$, but this is equaled by the calving loss, leading to an overall loss of $0.8\,\mathrm{m\,a^{-1}}$. The low slope of this glacier (1.6°) makes it difficult to reestablish equilibrium as it retreats. The bed of the glacier remains below sea level for at least 70% of its length. The glacier retreated 1300 m from 1990 to 2015.

6.4 Sorkappland

Vasilievbreen is a glacier that terminates on the east coast of the southern island of Svalbard, a short distance southeast of Hornsund (Fig. 6.7). In 1990, this glacier had a single continuous terminus margin along the coast, merging with Skilfonna on the northeast side. The glacier retreated $50\,\mathrm{m\,a^{-1}}$ from 1936 to 1990, as the embayment of Isbukta expanded (Blaszczyk, Jania, and Hagen, 2009). The glacier has since separated into distinct termini; on each image colored arrows indicate the same specific location – red arrow: a developing island named Fallknatten, yellow arrow: the tip of peninsula

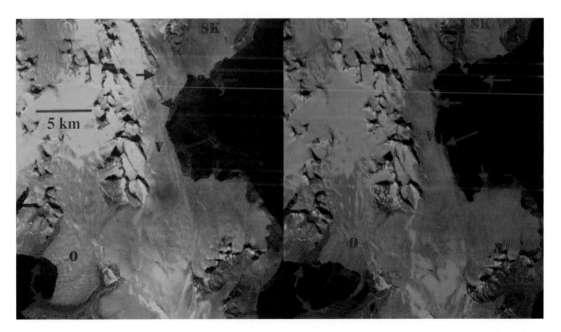

Figure 6.7 Outlet glaciers of Sorkappland: Olsokbreen (O), Vasilievbreen (V), and Skilfonna (SK) in 1990 (left) and 2014 (right) Landsat images. Red arrows indicate the same locations.

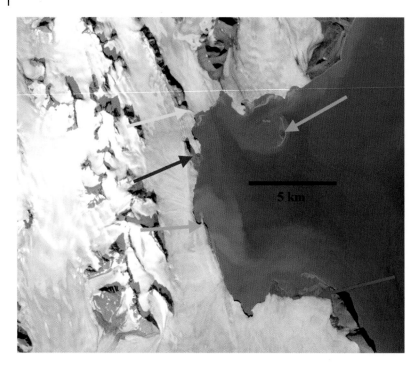

Figure 6.8 Google Earth image of the Sorkappland region. (Google Earth.)

called Gedenovfjellet, the orange arrow: an island called Morenetangen, the green arrow: an emerging island, and the purple arrow: an area of new coastline on the south side of Isbukta (Fig. 6.8).

In 1990, the terminus is continuous. The glacier reaches the coastline at the purple arrow. There is no evident land at the green arrow. The Skilfonna terminus approaches quite close to the island at the orange arrow, and there is a nunatak surrounded by ice at the red arrow. In 2014, there is no longer a continuous terminus. By 2014, the strip of land at the purple arrow has expanded to a length of 2 km and a width of 300–500 m, a retreat of 400 m. At the green arrow, a 3 km long rib of bedrock is exposed, and the glacier has retreated 500 m since 1990. At the red arrow, the nunatak of 1990 is now at the terminus of the glacier and is much larger, a retreat of 750 m. At the yellow arrow, the end of a peninsula is now much closer to the ocean, a retreat of 900 m. At the orange arrow, this moraine-based island has been eroding, but is also much further from the glacier, a retreat of Skilfonna of 800 m. The glacier now has five distinct terminus segments that can retreat independently of each other. That this retreat occurred along a front that is 20 km long represents a loss in glacier area of approximately 10 km^2. This is more significant than the actual distance of retreat. The snow line on the glacier in 2014 is above 300 m with the melt season still ongoing, which leaves most of the glacier in the melt zone.

Olsokbreen is the southernmost of the major glaciers on the west coast of Sorkappland. Olsokbreen has a 5 km calving front and its retreat was observed to have retreated 3.5 km from 1900 to 2008 (Blaszczyk, Jania, and Hagen, 2009). The glacier has pulled back from a peninsula extending into the sound from the north side of the fjord that the glacier ended upon in 1990 (red arrow), with a concave calving front. In 2014, the calving front has become almost west to east with a narrow coastal shelf fringe on the south. This indicates a much greater retreat of the south side than the north side. The retreat has been 4700 m on the south since 1990 and 200 m on the north side. The embayment appears

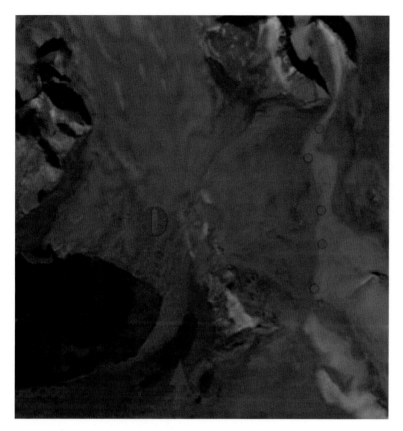

Figure 6.9 Landsat image of Olsokbreen in 2013, with the snow line back at the divide with Vasilievbreen. Orange arrow indicates new proglacial lake.

to narrow at this point with a ridge extending from the south side; this should slow the retreat of the southern margin. In 2013, the snow line reaches the divide with Vasilievbreen to the east (Fig. 6.9). Note the proglacial lake along the glacier margin, which was ice filled in 1990 (orange arrow).

References

Blaszczyk, M., Jania, J., and Hagen, J. (2009) Tidewater glaciers of Svalbard: recent changes and estimates of calving fluxes. *Polish Polar Research*, **30** (2), 85–142.

Blaszczyk, M., Jania, J., and Kolondra, L. (2013) Fluctuations of tidewater glaciers in Hornsund Fjord (Southern Svalbard) since the beginning of the 20th century. *Polish Polar Research*, **34** (4), 327–352. doi: 10.2478/popore-2013-0024

Kvamstø, N.G., Steinskog, D.J., Stephenson, D., and Tjøstheim, D. (2012) Estimation of trends in extreme melt-season duration at Svalbard. *International Journal of Climatology*, **32**, 2227–2239. doi: 10.1002/joc.3395

Luckman, A., Benn, D., Cottier, F. *et al.* (2015) Calving rates at tidewater glaciers vary strongly with ocean temperature. *Nature Communications*, **6**, 8566.

Nuth, C., Kohler, J., König, M. *et al.* (2013) Decadal changes from a multi-temporal glacier inventory of Svalbard. *The Cryosphere*, **7**, 1603–1621. doi: 10.5194/tc-7-1603-2013

Nuth, C., Moholdt, G., Kohler, J. *et al.* (2010) Svalbard glacier elevation changes and contribution to sea level rise. *Journal of Geophysical Research*, **115**, F01008. doi: 10.1029/2008JF001223

Oerlemans, J., Jania, J., and Kolondra, L. (2011) Application of a minimal glacier model to Hansbreen, Svalbard. *The Cryosphere*, **5**, 1–11. doi: 10.5194/tc-5-1-2011

Palli, A., Moore, J.C., Jania, J., and Głowacki, P. (2003) Glacier changes in southern Spitsbergen, Svalbard, 1901–2000. *Annals of Glaciology*, **37**, 219–225. doi: 10.3189/172756403781815573

Pętlicki, M., Ciepły, M., Jania, J. *et al.* (2015) Calving of a tidewater glacier driven by melting at the waterline. *Journal Glaciology*, **61** (229), 851–863.

Vieli, A., Jania, J., and Kolondra, L. (2002) The retreat of a tidewater glacier: observations and model calculations on Hansbreen, Spitsbergen. *Journal of Glaciology*, **48** (163), 592–600. doi: 10.3189/172756502781831089

Walczowski, W. and Piechura, J. (2011) Influence of the West Spitsbergen Current on the local climate. *International Journal of Climatology*, **31** (7), 1088–1093. doi: 10.1002/joc.2338

7

Novaya Zemlya

Overview

The glaciers of northern Novaya Zemlya, Russia, are truly generally out of sight out of mind, their remoteness and lack of importance as a water resource being the key reasons. It seems particularly important to pay attention to these glaciers due to the recent changes in sea ice cover that have left a much longer duration of open water around the island particularly to the west in the Barents Sea. Novaya Zemlya has glaciers draining east and west from the main spine of the island. The further north the more prevalent the tidewater glaciers are (Sharov, 2005). Here we examine glaciers along both the Kara and Barents Sea coasts of the island. The glaciers terminating in the Barents and Kara Seas have been retreating like all tidewater glaciers in northern Novaya Zemlya (LEGOS, 2006). From 1992 to 2010, Carr, Stokes, and Vieli (2013) identified an average retreat rate of $52\,\mathrm{m\,a^{-1}}$ for tidewater glaciers on Novaya Zemlya and $5\,\mathrm{m\,a^{-1}}$ for land-terminating glaciers. The retreat rate was more rapid for Barents Sea tidewater glaciers at $62\,\mathrm{m\,a^{-1}}$ than for the Kara Sea at $42\,\mathrm{m\,a^{-1}}$. The retreat rate increased during the 1992–2010 period after a period where half the termini were stable from 1963–1995 (Zeeberg and Forman, 2001). The beginning of rapid retreat coincides with the onset of rapid decline in summer sea ice concentration in the Kara Sea in 2003 (Fig. 7.1) (Perovich and Richter-Menge, 2015; Carr *et al.*, 2014), and Barents Sea summer sea ice concentration declines after 2000. In this chapter we focus on tidewater glaciers of northern Novaya Zemlya during the 1990–2015 period using Landsat image comparison.

7.1 Kropotkina Glacier

Kropotkina Glacier is a tidewater glacier on the southeast coast of Novaya Zemlya that drains into Vlaseva Bay. LEGOS (2006) indicated limited retreat from 1952 to 1988 and initiation of retreat with a loss of 2.3 km^2 of area from 1990 to 2000.

In 1990 the terminus is at the red arrow indicating a peninsula on the east side of the terminus (Fig. 7.2). The yellow arrow indicates a lake beyond an eastern terminus lobe with limited drainage down a river adjacent to the glacier (purple arrow). In 1998 there is a minor retreat of the main terminus on both the east and west sides. Little change is seen elsewhere. In 2015 a substantial embayment has developed above the red arrow. Retreat is limited on the western side of the terminus. The eastern terminus lobe has retreated as well, and the drainage channel adjacent to the glacier is less restricted leading to a less extensive lake. The lake is mostly filling the region occupied by ice 13 years before.

The eastern most terminus lobe is collapsing and is now surrounded by a lake (yellow arrow). The drainage river along the eastern margin from the impounded lake is no longer necessary, as there is

Recent Climate Change Impacts on Mountain Glaciers, First Edition. Mauri Pelto.
© 2017 John Wiley & Sons, Ltd. Published 2017 by John Wiley & Sons, Ltd.

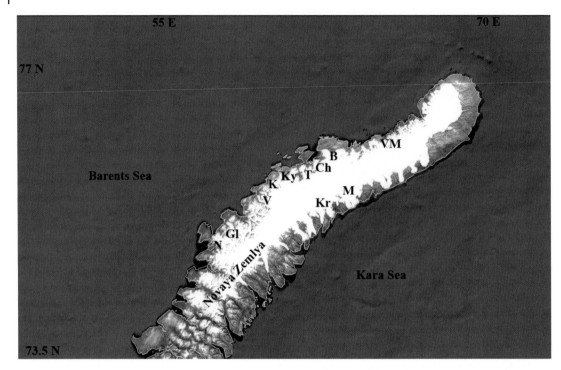

Figure 7.1 Novaya Zemlya viewed in Google Earth image with key glaciers identified. Kr = Kropotkina, M = Moshniy, N = Nizkiy, Gl = Glazova, V = Vilkitskogo, K = Krivosheina, Ky = Krayniy, T = Tasija, Ch = Chernysheva, B = Borzova, and VM = Mack and Velkena Glaciers. (Google Earth.)

lake connectivity at Vlaseva Bay. The main terminus to the east has retreated to the entrance to the lake for the eastern terminus lobe, which is a 2800 m retreat. The western side of the glacier remains aground on a peninsula but has receded 500 m. How long before this western part of the terminus to retreat into the expanding embayment? An area greater than 7 km^2 has transitioned from glacier ice to embayment in the last 25 years, almost all within the last 13 years. The retreat has mainly been via calving, and with an expanding calving front and reduced pinning points along the margin, the rapid retreat and area loss are not over (Fig. 7.3).

7.2 Moshniy Glacier

Moshniy Glacier is a tidewater glacier on the southeast coast of Novaya Zemlya that drains into the Kara Sea just northeast of Kropotkina (Fig. 7.2). In 1990 the glacier ends at the coast with a narrow coastal fringe extending short distance from the western side (red arrow). There the main terminus is slightly convex from the coast extending into the sea. The terminus had not changed appreciably from 1952 to 1988 but had lost 7.7 km^2 of area from 1990 to 2000 (LEGOS, 2006). In 2015 the fringe of coastline extends 2 km further across the glacier front from the west. The western margin has retreated 500–1000 m since 1990. The eastern terminus no longer extends to the coastline, as an embayment has developed. Retreat since 1990 has been 800 m. Like at Kropotkina, the embayment

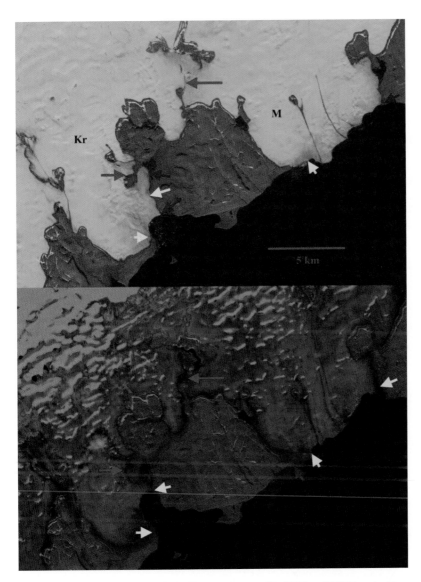

Figure 7.2 Kropotkina and Moshniy Glaciers compared in 1990 and 2015 Landsat images. Red arrows indicate 1990 terminus positions, yellow arrows 2015 terminus positions, and purple arrows upglacier thinning.

is beginning a rapid expansion that will lead to rapid retreat of the eastern section of the ice front. Upglacier thinning is evident on the western margin of the glacier also (purple arrow).

7.3 Vilkitskogo Glacier

Vilkitskogo Glacier has two termini that were nearly joined in Vilkitsky Bay in 1990 (Fig. 7.4). The north and south glaciers both terminated at the mouth of their respective fjords. In 2015 Vilkitskogo

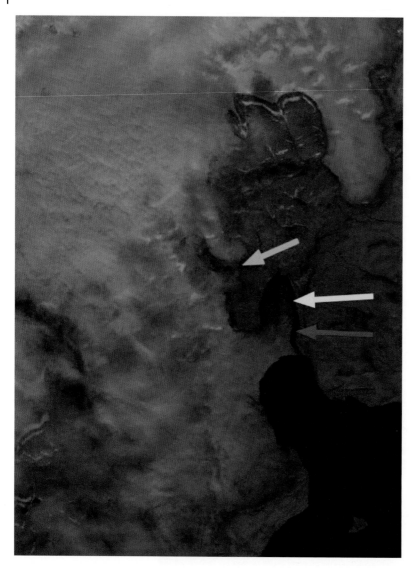

Figure 7.3 2014 Landsat image indicating the margin of Kropotkina. Red arrow indicates the 1990 terminus location; the purple arrow indicates what used to be a drainage river adjacent to the glacier, but is now simply a narrow lake connection. Yellow arrow indicates the 2015 margin of a collapsing arm of the terminus.

North has retreated 2500 m along the northern side of the fjord and 1500 m along the south side. This fjord has no evident pinning points, and the rapid calving retreat should continue. Vilkitskogo South has retreated 500 m on the west side and 1 km on the east side. The retreat has exposed a new island in the center of the glacier. The glacier is currently terminating on another island. Retreat from this pinning point will allow more rapid retreat to ensue. Upglacier thinning is evident in the expansion of bedrock areas and medial moraine width (purple arrows).

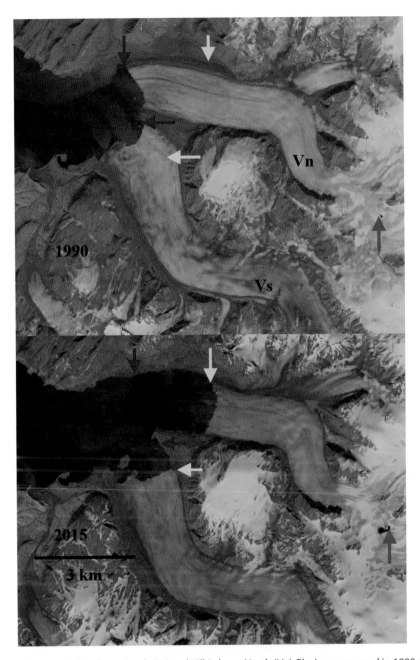

Figure 7.4 Vilkitskogo South (Vs) and Vilkitskogo North (Vn) Glaciers compared in 1990 and 2015 Landsat images. Red arrows indicate 1990 terminus positions, yellow arrows 2015 terminus positions, and purple arrows upglacier thinning.

7.4   Krivosheina Glacier

Krivosheina Glacier is a tidewater glacier on the northwest coast of Novaya Zemlya ending in the Barents Sea. LEGOS (2006) identify that Krivosheina Glacier lost 3.3 km^2 of area from 1990 to 2000. Here we examine satellite images from 1990 to 2015 (Fig. 7.5). In 1990 the glacier ended on an island (red arrow Point A on each image). The terminus extended due south and due east from the island in 1990. In 2009 a deep water channel was present between the island and the terminus. In 2015 the terminus had retreated 2000 m from its 1990 terminus position on the island and 1400 m along the southern margin of the glacier. Given that the glacier is 4.6 km wide, the retreat averaged 1.6 km, which equals an area loss of 7 km^2 since 1990. In 2013 the snow line is also at a higher elevation on the glacier. The glacier width remains similar for 2 km before it narrows and turns, which will likely act as a pinning point. Thinning upglacier is evident at purple arrow, indicating side valleys that could develop further proglacial lakes.

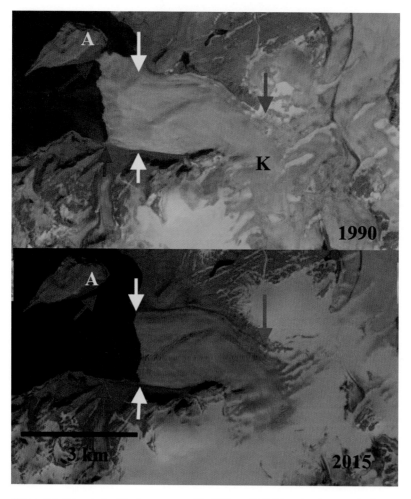

Figure 7.5 Krivosheina Glacier compared in 1990 and 2015 Landsat images. Red arrows indicate 1990 terminus positions, yellow arrows 2015 terminus positions, and purple arrows upglacier thinning. Point A indicates a new island that has formed.

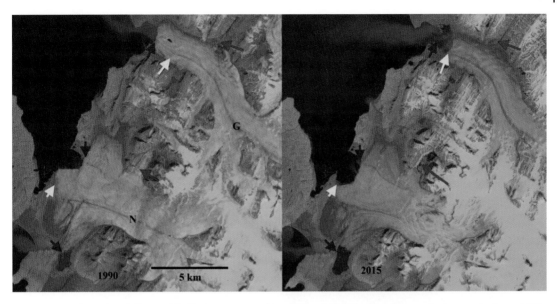

Figure 7.6 Nizkiy (N) and Glazova (G) Glaciers compared in 1990 and 2015 Landsat images. Red arrows indicate 1990 terminus positions, yellow arrows 2015 terminus positions, and purple arrows upglacier thinning.

7.5 Nizkiy Glacier

Nizkiy and Glazova Glaciers are on the west coast of the island. LEGOS (2006) identified each as losing 1.2 and 1.4 km^2 in area, respectively, from 1990 to 2000. Here we compare a Landsat image from 1990 and 2015 (Fig. 7.6).

Nizkiy is the southern glacier of the two. The Nizkiy Glacier has several termini in lakes and one in the Barents Sea. The main terminus juts north out to the end of a peninsula in 1990, with an embayment developing from the peninsula across to the northern edge of the terminus. The middle terminus ends in a proglacial lake, and in 1990 the terminus largely envelops an island in the lake. The southernmost terminus is in a proglacial lake, which is indicated by a red arrow as well. The 2015 Landsat image indicates the continued reduction in Nizkiy Glacier width reaching the peninsula at the yellow arrow, having retreated 1200 m from the 1990 position (Fig. 7.7). Hence, It will likely be quite soon when the proglacial lake with the island joins with the Barents Sea. The northern edge of the terminus has changed little, but the size of the embayment between the northern edge and the peninsula has doubled since 1990. The expansion of the proglacial lake east of the peninsula has exposed another new island (yellow arrow). The southernmost terminus has retreated expanding the proglacial lake (purple arrow).

7.6 Glazova Glacier

In 1990 Glazova Glacier terminates on a small peninsula near the entrance to a newly forming fjord. (Fig. 7.6). In 2015 Glazova Glacier has retreated 1500 m along the southern side of the fjord and 1100 m on the northern side. The fjord appears to widen beyond the current terminus, which will enhance calving and retreat.

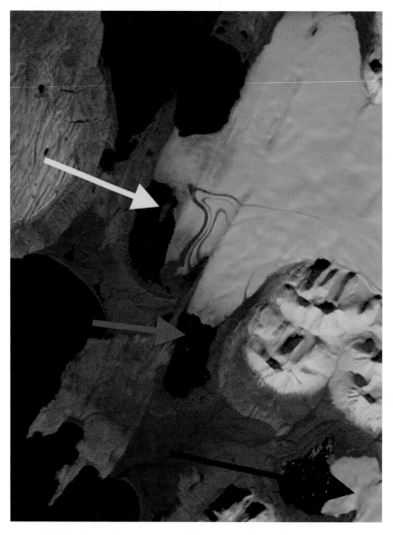

Figure 7.7 The Nizkiy Glacier proglacial lake system. The yellow arrow indicates a new island in the lake that will soon be joined to the Barents Sea.

7.7 Krayniy Glacier

Krayniy Glacier is an outlet glacier that drains the northern side of the Novaya Zemlya Ice Cap into the Barents Sea. This outlet glacier is just southwest of Tasija Glacier. The terminus of the glacier has a pinning point on an island at present (Fig. 7.8).

From 1990 to 2015 the glacier has retreated more on the eastern margin with 1250 of retreat opening up the embayment. Retreat at the island in the glacier center has been 500 m since 1990. The western section of the glacier has retreated little. The eastern embayment will continue to drive retreat and glacier thinning that will reduce contact with the island pinning the eastern half of the glacier. The thinning is evident at the purple arrows. The glacier will likely retreat from this island in a manner

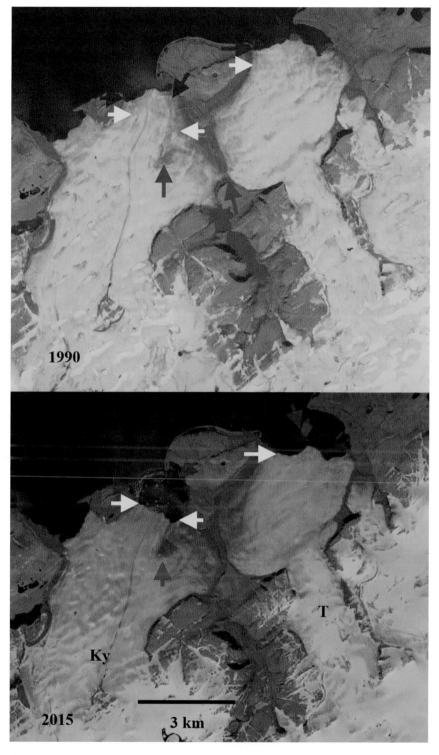

Figure 7.8 Tasija (T) and Krayniy (Ky) Glaciers compared in 1990 and 2015 Landsat images. Red arrows indicate 1990 terminus positions, yellow arrows 2015 terminus positions, and purple arrows upglacier thinning.

similar to that of Tasija and Chernysheva Glaciers, which will lead to increased rate of retreat of the entire ice front.

7.8 Taisija Glacier

Taisija Glacier is an outlet glacier that drains the northern side of the Novaya Zemlya Ice Cap into the Barents Sea. This outlet glacier is just southwest of Chernysheva Glacier, and similar to that, the glacier has retreated from an island since 1990.

Here we examine the glacier using Landsat images from 1990 and 2015 (Fig. 7.8). In 1990 the glacier terminus was grounded on an island near the center of the glacier (red arrow). The western margin red arrow is near the tip of a peninsula. On the east side an embayment exists in 1990 and the terminus is on a small island (red arrow). In 2006 the glacier center was still grounded on the island. An embayment has formed on the west side of the glacier and it has retreated from the peninsula. On the east side the glacier has retreated from the island, though this embayment has some sea ice in it that makes the retreat less evident. In 2015 the glacier in the center had retreated from the island, which acted as a pinning point. The embayments on the east and the west have both expanded. The retreat of this glacier is 1 km on the west, 1.4 km in the center, and 1.6 km on the west since 1990. The large mostly ice-filled embayment is now poised to collapse via calving and basal melting.

7.9 Chernysheva Glacier

Chernysheva Glacier is on the northwest coast of Novaya Zemlya. The glacier terminates in the Barents Sea and has been retreating like all tidewater glaciers in northern Novaya Zemlya (LEGOS, 2006). In this post we examine the changes using Landsat images from 1990 to 2015 (Fig. 7.9). In 1990 the

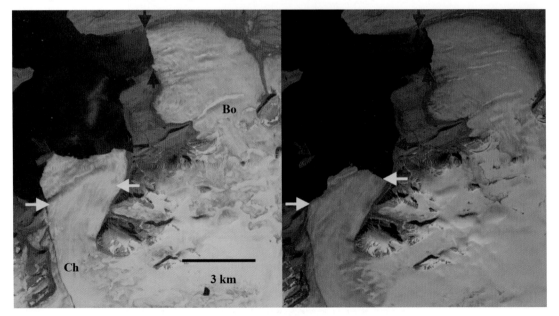

Figure 7.9 Chernysheva (Ch) and Borzova (Bo) Glaciers compared in 1990 and 2015 Landsat images. Red arrows indicate 1990 terminus positions, and yellow arrows indicate 2015 terminus position.

glacier ended on the island, which is a pinning point that would limit calving at the time (red arrow near glacier center). The glacier front then extended nearly due east. The 1990 image indicates the western margin at a subglacier bedrock ridge extending northeast from the yellow arrow 70% of the way across the glacier. In 2006 the glacier has pulled back from the island; there is, however, sea ice between the glacier front and the island. In 2015 the glacier retreat from this island is 3100 m. The retreat from the east margin is 2600 m. The bedrock ridge on the western margin has provided stability, and retreat has been minor. The eastern margin has retreated past this pinning point, and it appears that it will continue to retreat rapidly. This will eventually destabilize the western margin. Interestingly the embayment here is very similar to that of Nizkiy and Glazova Glaciers. The reduction in the duration of Barents Sea ice cover in front of the glacier has certainly helped to increase recent retreat, allowing access of both warmer water and wave action that both increase calving rates (Luckman *et al.,* 2015).

7.10 Borzova Glacier

Borzova Glacier is just east of Chernysheva Glacier and has a nearly north–south-oriented calving front. LEGOS (2006) indicated an expansion of this glacier from 1990 to 2000. From 1990 to 2015 there is minor retreat of the terminus that is less than 500 m. There is also little evidence of upglacier thinning. This glacier will begin to retreat faster if it retreats into deeper water, otherwise slow retreat will continue.

7.11 Mack and Velkena Glaciers

Mack and Velkena Glaciers are tidewater glaciers on the northwest coast of Novaya Zemlya that drain into Legzdina Gulf. The map shown in the following from LEGOS (2006) indicates that the glaciers joined in 1952 and 1976, and separated by less than 1 km in 1988 (Fig. 7.10). Here we examine Landsat images from 1988 to 2013 to identify changes in Mack and Velkena Glaciers.

   The arrows and letters in each image are at fixed locations: the yellow arrow indicates the peninsula where Velkena and Mack Glaciers separated. The red arrow indicates the 1988 eastern terminus of Mack Glacier, and the purple arrow indicates the 1988 position of the western terminus of Velkena Glacier. The green and pink Cs, respectively, indicate areas of intense crevassing in 2013. In 1988 the glaciers are separated by just 500 m adjacent to yellow arrow; the crevassing is limited at both Cs. In 2000 Velkena and Mack Glaciers are separated by 1.4 km; retreat of Mack from the red arrow is 500 m, and retreat of Velkena Glacier from the purple arrow is 600 m. In 2006 Mack Glacier has retreated 1000 m from the red arrow, and Velkena Glacier has retreated 1200 m from the purple arrow. Extensive crevassing is not yet evident at C.

   In 2011, at both locations, crevassing at C is readily apparent in the satellite imagery. Retreat has continued for both glaciers: 1250 m for Mack Glacier from red arrow and 1600 m for Velkena Glacier from purple arrow. In 2013 the extensive crevasses zone (C) for both glaciers is within 500 m of the ice front. The increase in crevassing is indicative of glacier acceleration; this is likely due to a steeper slope of the glacier near the current terminus and the proximity to the calving front. This also suggests that the retreat will remain rapid in the near future. Retreat of Mack Glacier from 1988 to 2013 has been 1800 m on the east side (red arrow) and 2000 m on the west side. On Velkena Glacier, the retreat has been 2800 m on the west side (purple arrow) and 1000 m on the east side. Both glaciers have been receding at a rate greater than the $50\,m\,a^{-1}$ noted as average by Carr, Stokes, and Vieli (2014).

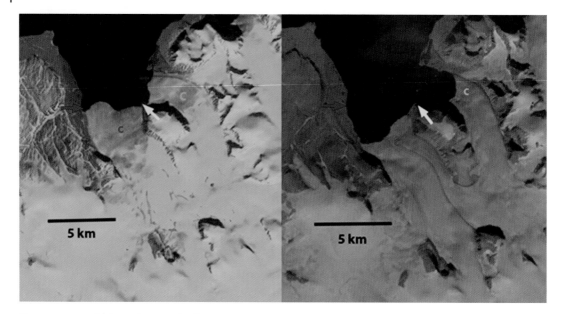

Figure 7.10 Mack Glacier (east) and Velkena Glacier (west) compared in 1988 and 2013 Landsat images.

References

Carr, J., Stokes, C., and Vieli, A. (2014) Recent retreat of major outlet glaciers on Novaya Zemlya, Russian Arctic, influenced by fjord geometry and sea-ice conditions. *Journal of Glaciology*, **60** (219), 155–170.

Carr, J., Stokes, C., and Vieli, A. (2013) Recent progress in understanding marine-terminating Arctic outlet glacier response to climatic and oceanic forcing. Twenty years of rapid change. *Progress in Physical Geography*, **37** (4), 436–467. doi: 10.1177/0309133313483163

Kouraev, A., Legrésy, B., and Remy, F. (2006) *Northern Novaya Zemlya Outlet Glaciers: 1990–2000 Changes.* LEGOS INTEGRAL Team, http://www.legos.obs-mip.fr/pdf/equipe/cryosphere-satellitaire/projets-spatiaux/northen-novaya-zembya-outlet (last accessed 27 May 2016).

Luckman, A., Benn, D., Cottier, F. *et al.* (2015) Calving rates at tidewater glaciers vary strongly with ocean temperature. *Nature Communications*, **6**, 8566.

Perovich, D.K., and Richter-Menge, J.A. (2015) Regional variability in sea ice melt in a changing Arctic. *Proceedings of the Royal Society*, **373**. doi: 10.1098/rsta.2014.0165.

Sharov, A. (2005) Studying changes of ice coasts in the European Arctic. *Geo-Marine Letters*, **25** (2–3), 153–166. doi: 10.1007/s00367-004-0197-7

Zeeberg, J. and Forman, S. (2001) Changes in glacier extent on north Novaya Zemlya in the twentieth century. *Holocene*, **11** (2), 161–175. doi: 10.1191/095968301676173261

8

North Cascade Range, Washington USA

Overview

The North Cascades, Washington, are host to more glaciers than any other region of the United States outside of Alaska. Post *et al.* (1971) identified 750 glaciers in the region. The glaciers are crucial to regional water resources for irrigation, hydropower, municipal supply, and aquatic life most notably salmon. The glaciers are small and occupy a temperate maritime climate setting, making them particularly sensitive to climate. In 1984 the North Cascade Glacier Climate Project (NCGCP) began monitoring glaciers across the mountain range to identify their response to climate change. The program has conducted annual surveys of glaciers for the last 32 years. Here we focus on results from glaciers in the Skykomish River Basin and on Mount Baker.

Analysis of key components of the alpine North Cascade hydrologic system indicates significant changes in glacier mass balance, terminus behavior, alpine snowpack, and alpine streamflow from 1950 to 2005 (Pelto, 2008; Luce *et al.*, 2014), which has continued up to 2014. Glacier runoff is of particular importance to streamflow and stream temperature and consequently aquatic life late in the summer when other water sources are at a minimum (Dery *et al.*, 2009; Pelto, 2008). Without consideration of changing glacier runoff, impacts of climate change on the Nooksack River cannot be assessed.

Climate observations in the Pacific Northwest, United States, show an accelerated warming for the 1970–2012 time periods of approximately 0.2 °C per decade (Abatzoglou, Rupp, and Mote, 2014). North Cascade temperatures have increased during the twentieth century. On average, the region warmed about 0.6 °C; warming was largest during winter (Abatzoglou, Rupp, and Mote, 2014).

The April 1 snow water equivalent (SWE) from six long-term USDA SNOTEL stations (Fish Lake, Lyman Lake, Park Creek, Rainy Pass, Stampede Pass, Stevens Pass) declined by 23% from the 1946–1976 period to the 1977–2014 period (Fig. 8.1). Winter season precipitation has declined only slightly by 3% at the SNOTEL stations and at Diablo Dam. Thus, most of the loss reflects increased melting of the snowpack or rain events during the winter season.

Mote, Hamlet, and Salathe (2008) referred to the ratio of 1 April SWE to November–March precipitation, as the storage efficiency. This ratio declined by 28% for the 1944–2006 period (Mote, Hamlet, and Salathe, 2008), which indicates that less of total precipitation is being retained as snowpack. Pelto (2008) noted a similar change specifically for the six long-term North Cascade SNOTEL stations, for the 1946–2014 period, which indicates a 22% decline in this ratio.

Recent Climate Change Impacts on Mountain Glaciers, First Edition. Mauri Pelto.
© 2017 John Wiley & Sons, Ltd. Published 2017 by John Wiley & Sons, Ltd.

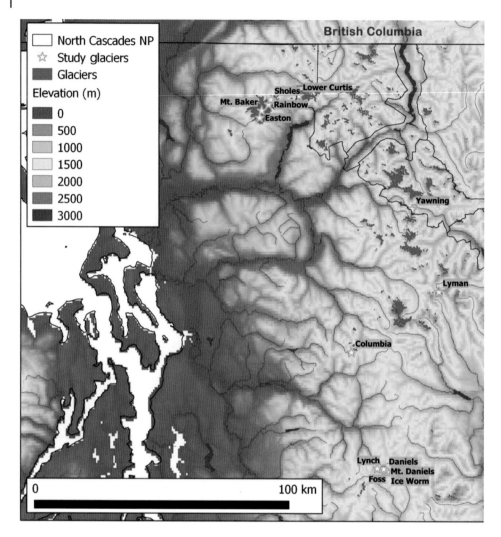

Figure 8.1 Map of the North Cascade region and key glacier areas observed (prepared by Ben Pelto).

8.1 Skykomish River Basin

Glacier retreat and changes in summer runoff have been pronounced in the Skykomish River Basin, North Cascades, Washington from 1950 to 2009 (Pelto, 2011). In the Skykomish River watershed, from 1958 to 2015, glacier area declined from 3.8 to 2.0 km². Columbia, Foss, Hinman, and Lynch Glaciers, the primary glaciers in the basin, declined in area by 10%, 60%, 90%, and 35%, respectively, since 1958. Annual mass balance measurements completed from 1984 to 2015 on Columbia, Foss, and Lynch Glaciers indicate a mass loss of 17 m w.e.

8.1.1 Lynch Glacier

It is on the north side of Mount Daniels, the highest point in the Skykomish Watershed, and drains into the South Fork Skykomish River. In the United States Geological Survey (USGS) maps of the

Figure 8.2 Lynch Glacier 1996; note the lack of bedrock in the upper right.

region, the glacier flows down the mountain ending in basin, with a modest fringe of water showing. This lake, usually referred to as Pea Soup Lake, expanded rapidly between 1978 and 1983, as the portion of the glacier occupying this basin disintegrated; in 1988, the lake is fully open water (Fig. 8.2). Lynch Glacier retreated 390 m from 1950 to 1979, almost all of it occurring in a breakup of the glacier in Pea Soup Lake. From 1979 to 2015, the glacier has retreated 150 m from the lakeshore. Annual mass balance measurements indicate the loss of 16 m of ice thickness on average. More importantly in 2003, on the upper west section of the glacier section, a bedrock ridge and scattered outcrops were exposed. The features have continued to expand, indicative of thinning of the glacier in its accumulation zone; note the rock outcrop on the upper right portion of the glacier in 2015 in Fig. 8.3. This is an indicator of a glacier that cannot survive current climate. Lynch Glacier has lost 35% of its area since 1958. As Pea Soup Lake has matured, it has turned from pea green to turquoise in color.

8.1.2 Hinman Glacier

Hinman Glacier is on the west side of Mount Hinman and drains into the South Fork Skykomish River. This was the largest glacier in the North Cascades south of Glacier Peak 50 years ago. Today, it is nearly gone. Hinman Lake, unofficial name, has taken the place of the lower portion of the former glacier, which still has a couple of separated relict ice masses. In the USGS map based on 1958 photographs, the glacier extends from the top of Mount Hinman at 2300 m to the bottom of the valley at 1525 m. Figure 8.4a is of Hinman Glacier from the west in 1988, the Hinman Glacier is now a group of four separated ice masses; three are significant in size still. Figure 8.4b is a 2009 view from the far end, north end of Lake Hinman up the valley and mountain side that was covered by the Hinman Glacier, now 90% gone compared to 1958. The new lake is 1 km long. This is no longer a glacier and is just a few relict pieces of ice as viewed from the 1958 terminus location in 2009; the largest has an area of 0.05 km^2.

Figure 8.3 Lynch Glacier in 2015; bedrock rib is visible in upper right of the glacier.

(a)   (b)

Figure 8.4 (a) Outline of Hinman Glacier in 1988 and (b) a view from the new Hinman Lake in 2009, as taken from the 1958 terminus location.

(a)   (b)

Figure 8.5 Foss Glacier in (a) 1988 and (b) 2015 taken from the west ridge of the Lynch Glacier. In 2015, the glacier is fragmenting into many small parts.

8.1.3 Foss Glacier

Foss Glacier is on the northeast side of Mount Hinman and drains into the South Fork Skykomish River. As recently as 1988, this glacier covered the majority of this mountain face (Fig. 8.5). Today, the glacier is rapidly thinning, separating into smaller parts, and retreating. Foss Glacier had by the middle of August lost all of its snow cover in 1992, 1993, 1994, 1998, 2003, 2005, 2009, 2014, and 2015. This has led to the thinning of the upper reaches of the glacier. Thinning of the upper reaches of a glacier is an indicator of a glacier that cannot survive current climate. The lower section detached from the upper section in 2003 and melted away in 2009. In 2015, the glacier has fragmented into four parts and will continue to melt away. Annual balance measurements indicate a loss of over 15 m of average ice thickness, which for a glacier that averaged 30–40 m in thickness represents 40–50% of the volume of the glacier lost between 1984 and 2009. The reduction in glacier extent has been 40% between 1984 and 2009 and 60% between 1958 and 2009.

8.1.4 Columbia Glacier

Columbia Glacier occupies a deep cirque above Blanca Lake ranging in altitude from 1460 to 1720 m and is the headwaters of the North Fork of the Skykomish River. Kyes, Monte Cristo, and Columbia Peaks surround the glacier with summits over 2200 m. This glacier has the lowest mean altitude of any substantial glacier in the North Cascades. The glacier is the beneficiary of locally heavy precipitation and orographic lifting over the surrounding peaks causing cooling of the air mass greater than that expected from the elevation of the glacier. Facing southeast Columbia Glacier is protected from afternoon sun. During the winters, storm winds sweep from the west across Monte Cristo Pass dropping snow in the lee on Columbia Glacier. Avalanches spilling from the mountains above descend onto and spread across Columbia Glacier. The avalanche fans created by the settled avalanche snows are 20 ft deep even late in the summer. Columbia Glacier has retreated 134 m since 1984 (Fig. 8.6). The head of the glacier has retreated 90 m. Lateral reduction in glacier width of 95 m in the lower section of the glacier and the reduction in glacier thickness are even more substantial as a percentage. The major issue is that the glacier is thinning as appreciably in the accumulation zone in the upper cirque basin as at the terminus. This indicates a glacier that is in disequilibrium with current climate and will melt away with a continuation of the current climate conditions, illustrated by Fig. 8.7. The glacier has lost 15 m in thickness since 1984 but still remains a thick glacier over 75 m in the upper basin and

Figure 8.6 Terminus comparison of Columbia Glacier – 1992 and 2015.

Figure 8.7 Painting of Columbia Glacier in the future. (Courtesy of Jill Pelto.)

will not disappear quickly. In 2003, 2004, 2005, and 2009, there has been nearly a complete exposure of the glacier surface and the loss of all firn and snow cover on the glacier. This exposed more than 50 annual layers in the 2005 image (Fig. 8.8). Streamflow and ablation are measured each summer on Columbia Glacier. Glacier runoff ranges from 15 to 30 cfs during the summer melt season from this glacier alone.

Figure 8.8 Columbia Glacier from its head in 2005 with almost the entire glacier exposed at the start of August; there was no retained snowpack by the end of the summer melt season. This has become a more common sight with 2015 having no retained accumulation.

8.1.5 Skykomish Streamflow Impact

An analysis comparing USGS streamflow records from the 1950–1985 to the 1985–2009 periods indicates that during the recent period, the Skykomish River summer streamflow (July–September) declined 26% in the watershed, spring runoff (April–June) declined 6%, while winter runoff (November–March) increased 10%. The minimum mean monthly August discharge from 1928 to 2015 occurred in 2015, 2003, and 2005 when streamflow was $11.8 \, \mathrm{m^3 \, s^{-1}}$, $15.1 \, \mathrm{m^3 \, s^{-1}}$, and $15.2 \, \mathrm{m^3 \, s^{-1}}$, respectively. From 1929 to 1985, streamflow was less than $14 \, \mathrm{m^3 \, s^{-1}}$ during the glacier melt season on a single day in 1951. From 1986 to 2015, there were 264 days with discharge below $14 \, \mathrm{m^3 \, s^{-1}}$, with 11 periods lasting for 10 consecutive days. Despite 15% higher ablation rates during the 1984–2009 period, the 45% reduction in glacier area led to a 35–38% reduction in glacier runoff between 1958 and 2009 (Pelto, 2011).

The 38% reduction in glacier runoff did not lead to a significant decline in the percentage summer runoff contributed by glaciers under average conditions; the contribution has remained in the range of 1–3% from July to September. The glacier runoff decline impacted river discharge significantly only during low-flow periods in August and September. In August 2003 and 2005, glacier ablation contributed $1.5–1.6 \, \mathrm{m^3 \, s^{-1}}$ to total discharge or 10–11% of August discharge. While declining glacier area in the region has and will lead to reduced glacier runoff and reduced late summer streamflow, it has limited impact on the Skykomish River except during periods of critically low flow, below $14 \, \mathrm{m^3 \, s^{-1}}$, when glaciers currently contribute more than 10% of the streamflow (Tennant, 1976).

8.2 Mount Baker and Nooksack River

Mount Baker, a stratovolcano, is the highest mountain in the North Cascade Range at 3286 m. Mount Baker has the largest contiguous network of glaciers in the range, with 12 significant glaciers covering 38.6 km^2 and ranging in elevation from 1320 to 3250 m (Fig. 8.9). Since 2012 I have partnered in a glacier runoff study with the Nooksack tribe, which refers to the mountains as Komo Kulshan, the great white (smoking) watcher. Kulshan watches over the Nooksack River Watershed, and its flanks are the main headwater sources for two of three branches of this river as well as the Baker River.

The Nooksack River consists of the North, South, and Middle Fork, which combine near Deming to create the main stem Nooksack River. Glaciers on the north and west side of Mount Baker drain into the Middle Fork and North Fork, while the South Fork has no glaciers. The Nooksack River empties into Puget Sound, part of the Salish Sea near Bellingham, Washington. Glaciers on the south and east side of Mount Baker drain into the Baker River that flows into the Skagit River at Concrete, Washington. The glaciers of Mount Baker are a key water resource to the aquatic life and the local communities and globally have proven to be some of the most sensitive to climate, responding quickly

Figure 8.9 Landsat 8 satellite image of Mount Baker. Light blue is snow, purple is bare glacier ice, dark green is evergreen forests, pink is a mixture of rock and alpine vegetation, and light green is areas dominated by deciduous shrubs.

to changes. This prompted initiation of a long-term glacier study by the NCGCP in 1984, which continues today, examining the response of Mount Baker glaciers and their resultant runoff to climate change.

The glaciers of Mount Baker have proven particularly quick to respond to climate change with terminus advance commencing within 5–15 years of a cooler wetter climate, as occurred in 1944, and terminus retreat within 3–12 years of a change to a warmer drier climate, as occurred in 1977 (Pelto and Hedlund, 2001).

On Mount Baker, all 12 glaciers advanced during the 1944–1978 period ranging from 60 to 750 m, an average of 480 m, and ended in 1978 (Heikkinen, 1984; Harper, 1993; Pelto, 1993). The first glaciers to advance were the glaciers with a steeper slope profile such as Coleman, Deming, and Roosevelt, which had all begun to advance in 1950 (Bengston, 1956; Long, 1956; Hubley, 1956). The last glaciers to advance were the glaciers with lower slope profile, Sholes and Easton Glaciers, which were still in retreat in 1952 (Long, 1956). This is an indication of the slower response time to a climate change of a glacier with a lower slope and slower velocity. In each case, the initial response of the glacier to climate change occurred within a decade.

Recent climate change has caused ubiquitous retreat of Pacific Northwest glaciers (Pelto, 1993; Key *et al.*, 2002; Granshaw and Fountain, 2006). In 1984, all the Mount Baker glaciers, which were advancing in 1970s, were again retreating (Harper, 1993; Pelto, 1993). The retreat on each Mount Baker glacier was measured in the field from benchmarks established in 1984 or 1985 at their recent maximum position (late 1970s to early 1980s). In each case, a maximum advance moraine had been emplaced. The retreat has continued to 2015 with the average retreat of 430 m for the nine principal Mount Baker glaciers in this interval. In Table 8.1, each Mount Baker glacier is described specifically, going counterclockwise from Deming Glacier.

8.2.1 Sholes Glacier

Sholes Glacier is at the headwaters of Wells Creek and has been the site of annual mass balance monitoring since 1990. From 1985 to 2003 retreat was limited (35 m), with the terminus often snow

Table 8.1 Characteristics and terminus change of Mount Baker glaciers.

Name	Area (km²)	Terminus (m)	Top (m)	Slope	Orientation	Terminus Change		
						LIAM–1947	1947–1979	1979–2014
Deming	4.79	1270	3270	0.39	215	−2700	+600	−615
Easton	2.87	1680	2900	0.35	195	−2420	610	−320
Squak	1.55	1700	3000	0.45	155	−2550	310	−300
Talum	2.15	1800	3000	0.55	140	−1975	280	−240
Boulder	3.46	1530	3270	0.50	110	−2560	740	−520
Park	5.17	1385	3270	0.41	110			−360
Rainbow	2.03	1340	2200	0.37	90	−1370	510	−480
Sholes	0.94	1610	2110	0.30	330	−1170	−60	−95
Mazama	4.96	1480	2980	0.44	10	−2500	−450	−410
Coleman		1350	3270	0.47	320	−2600	+400	−480
Roosevelt		1560	3270	0.45	320	−3200	+375	−400

Figure 8.10 View from the terminus of Sholes Glacier in 2015 to the mapped 1990 terminus position. Total retreat is 150 m.

covered into late summer. The glacier was thinning significantly during this period. This retreat has accelerated particularly during 2013–2015. Total retreat from 1985 to 2015 has been 150 m, with 20 m of retreat in 2015 alone (Fig. 8.10).

A stream level recorder has been installed during the 2013–2015 melt seasons to monitor glacier discharge 100 m from the glacier terminus. The standard mass balance measurement locations are indicated by blue dots, and the stream gaging station is at the black arrow (Fig. 8.11). Sholes Glacier is a broad slope glacier that lacks the extensive crevassing of other glaciers on Mount Baker and is not on the immediate slope of the volcano.

Each year, we assess snowpack distribution in midsummer and at the end of the summer. The change provides a direct measure of glacier runoff. The rate of melt can be tracked using the transient snowline (TSL). As the TSL rises, it traverses areas of known snow depth, providing a direct measure of ablation between the date of snow depth observation and TSL traversing the location (Fig. 8.12). The migration of the TSL can be identified in Landsat images as well as field observations (Fig. 8.13).

8.2.2 Rainbow Glacier

Rainbow Glacier is on the northeast side of Mount Baker; it descends from 2200 to 1340 m. This is the lowest starting elevation of the valley glaciers on the slopes of Mount Baker. The glacier is the headwater of Rainbow Creek, which drains into Baker Lake. The glacier currently terminates at

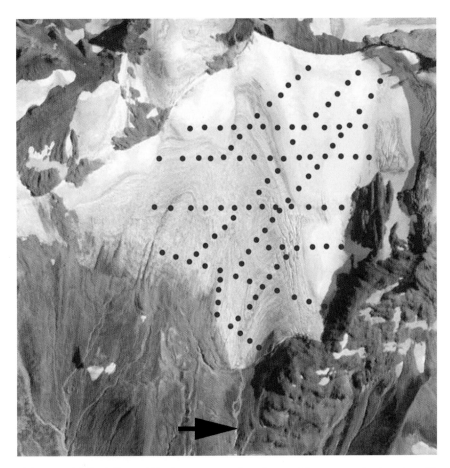

Figure 8.11 Sholes Glacier measurement network and potential stream gaging locations. Blue dots are accumulation measurement sites and in some instances ablation stake locations.

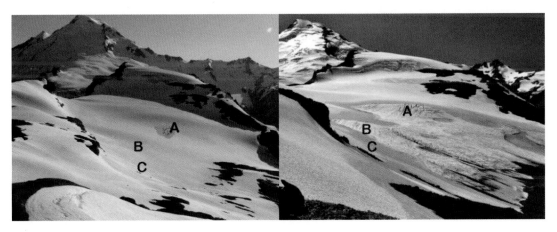

Figure 8.12 Comparison of snowpack on Sholes Glacier on August 4 and September 1, 2013.

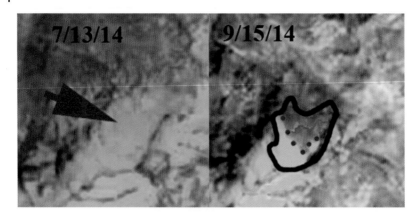

Figure 8.13 Comparison of snow extent in Landsat images from July 13 to September 15, 2014, indicating snow line position on Sholes Glacier.

Figure 8.14 Rainbow Glacier from Rainbow Ridge in September 2015. The saddle with Mazama Glacier is on skyline. (Courtesy of Tom Hammond.)

1350 m. The glacier has a fairly uniform slope of 0.29 from 1350 to 1800 m; a steeper icefall then leads up to the saddle where the glacier has only a minor slope (Fig. 8.14). In 1979 the glacier was advancing, building a terminal advance moraine, and had a crevassed terminus. In 1984 at the time of the first field survey of the glacier terminus, the glacier was still in contact with the advance moraine at 1175 m. The terminus was still actively crevassed and convex. In 1997 the terminus had retreated 225 m, no

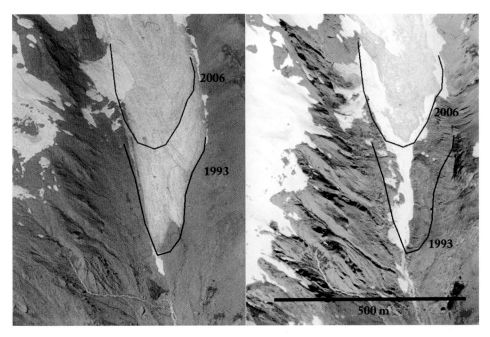

Figure 8.15 Digital Globe view of the terminus of Rainbow Glacier in 1993 and 2006, with the 1993 terminus in black and the 2006 terminus in red.

crevassing was evident in the terminus region, and the profile in the lower several hundred meters of the glacier was concave. A period of rapid retreat ensued until 2006. A comparison of Digital Globe images from 1993 and 2006 indicates the retreat during this period (Figs 8.15 and 8.16). Since 2006 the terminus has typically been covered by avalanche debris even late in the summer, slowing retreat, while the glacier above the terminus reach continues to thin. In 2013, 2014, and 2015, exceptional summer melt exposed 300 m of the glacier; the terminus has a concave profile and is uncrevassed indicating limited glacier movement, which will allow continued retreat in the near future. The overall glacier velocity has declined as evidenced by the reduction in crevassing. This glacier is the site of an annual balance monitoring program since 1984. From 1984 to 2014 the mean annual balance has been -0.34 m a^{-1}, which is a cumulative loss of -10.55 m or 12 m of glacier thickness.

8.2.3 Roosevelt Glacier

Roosevelt and Coleman Glaciers drain the northwest side of Mount Baker and are joined in the accumulation zone. Roosevelt Glacier retreated 3200 m from a joint terminus with Coleman Glacier at its Little Ice Age Maximum (LIAM) to 1949 (Heikkinen, 1984). Harper (1993) noted that the glacier advanced 350–400 m from 1949 to 1970. In 1984 the glacier was nearly in contact with its advanced moraine at 1400 m, which was ice cored and quite prominent. In 1993 the glacier had retreated 140 m. From 1993 to 2014 the glacier retreated above two lava flow bands and in 2014 terminated at 1560 m (Fig. 8.17). The retreat from 1984 to 2014 is 400 m, with the terminus now back to the 1949 terminus position (Fig. 8.18). The glacier is fed by three principal accumulation zones: (i) a glacier tongue that descends from the summit plateau at 3200 m, (ii) an avalanche-fed and direct snowfall region beneath the north ridge at 2200 m, and (iii) an avalanche and direct snowfall-fed region beneath the northwest face, at 2400 m. The annual snow line has averaged 2150 m on Roosevelt

Figure 8.16 Rainbow Glacier terminus in 2014 from Rainbow Ridge, indicating the 1984 terminus position as well. (Courtesy of Tom Hammond.)

Figure 8.17 Roosevelt Glacier terminus from Heliotrope Ridge, just reaching over the edge of the cliff in 2014. Coleman Glacier is in the foreground.

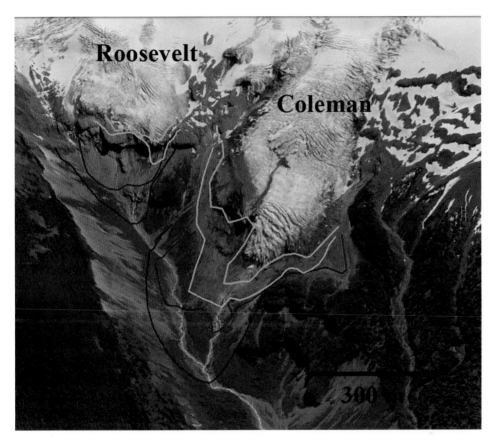

Figure 8.18 A 2009 Google Earth image of Roosevelt Glacier terminus with the purple line marking the 1979 terminus, red line the 1993 terminus, yellow line the 2003 terminus, and green line marking the 2009 terminus. (Google Earth.)

Glacier from 1984 to 2010. The lower portion of the glacier is thin, indicating that the retreat will continue.

8.2.4 Coleman Glacier

Coleman Glacier drains the northwest side of Mount Baker and is joined with the Roosevelt Glacier in its accumulation zone. The glacier drains the largest area of the volcano above 3000 m, which combined with its steep slope leads to a higher velocity and more crevassing than on other glaciers. The glacier retreated 2600 m from a joint terminus with the Roosevelt Glacier at its LIAM to 1949 (Heikkinen, 1984). The glacier separated from the Roosevelt Glacier in the 1930s. In 1949 the glacier had reached a minimum position (Bengston, 1956) before advancing 350 m by 1979 (Harrison, 1960; Harper, 1993). The advance terminated in the bottom of the Glacier Creek valley. From 1979 to 2014 the glacier has retreated up the west side of the valley wall, 480 m (Figs 8.18–8.20). The most rapid period of retreat was from 1979 to 1998. The current terminus descends over a steep lava flow and has active crevassing, indicating slow retreat in the near future. This glacier is currently the most active in the terminus region. The extensive melting in 2013–2015 may change this.

Figure 8.19 Coleman Glacier from Glacier Creek Road in 1984 still reaching the Glacier Creek valley bottom.

8.2.5 Deming Glacier

Deming Glacier is the headwaters of the Middle Fork Nooksack River. The glacier descends the southwest flank of Mount Baker beneath the Black Buttes. The glacier has the most spectacular icefall in the North Cascades from 2100 to 1700 m. The LIAM maximum moraine is at 970 m. Long (1953) noted that this glacier retreated 1360 m from 1907 to 1947, a total retreat of 2700 m from the LIAM. The glacier then experienced a period of advance. The glacier was still in contact with its 1970s advance moraine in 1985 during our first visit to the terminus. From 1979 to 2014 the glacier retreated 615 m; it is now close to the 1947 minimum position at 1250 m. A key change in the terminus section has been the expansion of the debris cover across the width of the glacier from 2003 to 2014. The lower 800 m of the glacier has a low slope and limited crevassing, suggesting that this section will be lost to retreat with current climate. The reduction in width and crevassing in the Deming Glacier Icefall indicates a reduced flow into the terminus reach of the Deming Glacier.

Each year since 1990, we have been able to observe the terminus of the Deming Glacier from our survey point, but as recent landslides indicate, visiting the terminus is too dangerous. Only in 3 years have we actually visited the terminus: 1985, 1996, and 2002. The first image is a 1979 Austin Post USGS image from 1979 when the glacier is in contact with both terminal and lateral moraines that are freshly built (Fig. 8.21). Figure 8.22 is from Google Earth showing in order the 1984 map position (blue), 1994 terminus (magenta), 2006 terminus (green), and 2011 terminus (yellow). The glacier retreated 160 m

Figure 8.20 Coleman (right) and Roosevelt Glaciers (left) in 2014 from Glacier Creek Road. The Glacier Creek valley bottom that Coleman Glacier reached in 1979 is evident in the middle foreground.

Figure 8.21 Deming Glacier in 1979, with red arrows indicating the location of a moraine that is developing.

Figure 8.22 Deming Glacier in a 2011 Google Earth image indicating the terminus location of 1984 (blue line), 1994 (magenta line), 2006 (green line), and 2011 (yellow line). (Google Earth.)

from 1984 to 1994, 16 m a^{-1}. From 1994 to 2006, the glacier retreated 240 m, 20 m a^{-1}. From 2006 to 2011, the glacier retreated 120 m, a rate of 24 m a^{-1}. The rate is still on the increase.

The sequence of images from our survey point begins with a view from 2003 showing the end of the glacier (Fig. 8.23). The front was still steep at the time and the width of the debris near the terminus is limited; clean ice width is 200 m. In 2008 the terminus is not as steep and most of the glacier width is debris covered. The last image is from 2014 indicating that the debris-free ice section is 40 m wide (Fig. 8.24). What is causing the narrowing of the debris-free ice is the reduction in velocity, the increased thinning of the clean ice in the center compared with the insulated debris-covered

Figure 8.23 Deming Glacier from the survey Point in 2003, indicating the 1985 terminus position.

ice at the edges. As the elevation difference increases, debris slides off the side of the developing debris-covered ridge (Pelto, 2000).

8.2.6 Easton Glacier

Easton Glacier flows down the south side of Mount Baker. The glacier terminates in a valley confined by lateral moraines that were built during the Little Ice Age, Railroad Grade to the west and Metcalf Moraine to the east. Easton Glacier extends from the slopes near Sherman Crater at 2950 m to the terminus at 1700 m, below the Railroad Grade and Metcalf Moraine. Each summer since 1990, NCGCP has measured the mass balance of this glacier. Snowpack typically increases from the terminus to 2500 m and then remains comparatively constant. In 1907 the glacier ended at 1250 m; by 1947, the glacier had retreated 2100 m (Long, 1953). Changes from 1911 to 2011 indicate the large change in ice thickness and extent at the terminus and limited change in the upper reach (Pelto and Brown, 2012). The Easton Glacier has a lower slope than the other largest glaciers on Mount Baker leading to a slower response to climate change (Pelto and Hedlund, 2001). The glacier started advancing after 1954 the last of the large Mount Baker glaciers to advance. The glacier advanced 500–600 m in 1979. The glacier was in contact with the moraine emplaced by this advance until 1990 (Fig. 8.25). The retreat was the last to begin of the large Mount Baker glaciers, due to the slower response time (Pelto and Hedlund, 2001). In 2014, the glacier had retreated 320–340 m from the advance moraine, $15 \, \text{m a}^{-1}$ (Figs 8.26–8.28). During this same period, the glacier has had a mean annual balance of $-0.51 \, \text{m a}^{-1}$, a cumulative loss of -12.7 m. This is equivalent to losing a 14 m of thickness. Given a thickness in 1990 between 60 and 75 m, this is about 20% of the total glacier volume (Harper, 1993;

Figure 8.24 Terminus in 2014 from survey point, indicating expansion of debris cover across terminus.

Finn *et al.*, 2012). The lowest 350 m of the glacier has limited crevassing and movement, indicating that retreat will continue. The glacier has developed a separate terminus on the east side at 1800 m. Each summer, we complete a profile across the glacier at 1950 m; the glacier has thinned 16 m across this profile since 1984 (Fig. 8.29).

8.2.7 Boulder Glacier

Boulder Glacier is the most prominent east side glacier on Mount Baker. This steep glacier responds quickly to climate change, and after retreating more than 2 km from its LIAM, it began to advance in the 1950s, as observed by William Long (1955, 1956). The glacier advance had ceased in 1979. From 1988 to 2008, we visited this glacier every 5 years, recording its changes. In 1988, the glacier had retreated only 25 m from its furthest advance of the 1950–1979 period. In 1993 the glacier had retreated 100 m from this position. At this time the lower 500 m of the glacier was stagnant. In 2003 the glacier had retreated an additional 300 m (Figs 8.30 and 8.31). In 2008 the glacier had retreated 490 m from its 1980 advance position, a rate of 16 m per year. The glacier as seen in 2008, despite the

Figure 8.25 Terminus of Easton Glacier from base camp in 1990. The advance moraine is the center of the image.

steep slope, has few crevasses in the debris covered lower 400 m of the glacier. This indicates that this section of the glacier is stagnant and will continue to melt away. The transition to active ice is at the base of the icefall on the right north side of the glacier. This glacier after 25 years of retreat is still not approaching equilibrium and will continue to retreat. This is a reflection of continued negative mass balance as measured on the adjacent Easton Glacier. It does respond fast to climate change, and the climate has not been good for this glacier. The glacier does have a consistent accumulation zone and can survive current climate. A Google Earth image from 2009 shows additional retreat now at 515 m from 1984 to 2009, 20 m a^{-1} (Fig. 8.32; green = 2009, brown = 2006, purple = 1993, yellow = 1984).

8.3 Glacier Runoff Impact

In 2013, The Nooksack River Watershed had glaciers with a combined area of 17.4 km^2, and the Baker River watershed had glaciers with a combined area of 29.6 km^2. In a typical summer (June–September), there is 3.2 m w.e. of ablation from glaciers in these watersheds (Pelto, 2008). This is equivalent to (150.4 million m^3) a mean summer discharge of 4.0 m^3 s^{-1} in the Nooksack River and 6.7 m^3 s^{-1} for Baker River (Pelto, 2015). In this region dominated by a temperate maritime climate, summer is the driest season, with limited precipitation from July 1 to October 1. This is the season when glaciers are especially important to regional water supplies (Fountain and Tangborn, 1985). The dependable supply of water in part due to the glaciers has led to the development of

Figure 8.26 Easton Glacier in 1998 from base camp. Terminal moraine is the ridge on the right foreground.

Figure 8.27 Easton Glacier terminus in 2003 from the base camp, with 1985 terminus indicated.

Figure 8.28 Easton Glacier from base camp in 2014.

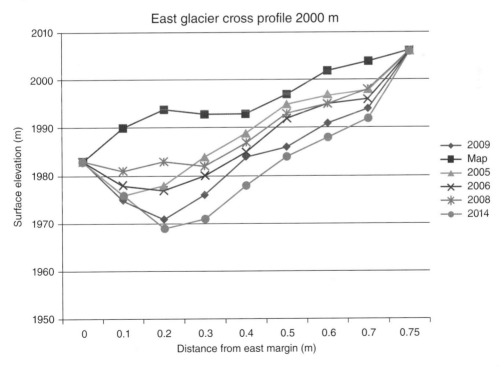

Figure 8.29 Change in surface elevation along a fixed profile across the glacier at approximately 2000 m. The map is based on a 1984 surface elevation.

Figure 8.30 Boulder Glacier in 2003 indicating the 1985 terminus position of an earlier survey.

Figure 8.31 On the Boulder Glacier terminus in 2003, the view beyond the margin indicates the recent retreat to the trimline of vegetation.

Figure 8.32 Boulder Glacier in Google Earth image from 1993 and 2009 indicating the retreat. (Google Earth.)

economies that rely on this water resource, from farming to sport and commercial fisheries to recreational activities and hydropower generation.

The Nooksack River system is home to five species of Pacific salmon – chinook, coho, pink, chum, and sockeye – as well as steelhead and bull trout. Three different fish species in the Nooksack River system have been listed as threatened under the Endangered Species Act (ESA), which include Chinook salmon, steelhead, and bull trout (Grah and Beaulieu, 2013).

Agricultural production, supported by Nooksack River water, has an approximate $290 million market value (Whatcom Farm Friends, 2014). Whatcom County's berry industry is world-class, raising the largest per capita crop of red raspberries in the world with a harvested area of over 29 km^2, supplying more than 65% of all the raspberries in the United States (Whatcom Farm Friends, 2014).

The City of Bellingham obtains 10% of its drinking water from the Middle Fork Nooksack River, which is fed by Deming and Thunder Glaciers on Mount Baker.

At Nooksack Falls, there is a hydropower plant constructed in 1906, on the North Fork Nooksack River that is rated at a production of 3.5 MW, though its operation has been sporadic since 1997. On the Baker River, there are two large hydropower projects that can generate 215 MW of electricity at the Upper and Lower Baker Dams. The Lower Baker Dam is located on the Baker River in Concrete, Washington. It is 87 m tall and the reservoir capacity of Lake Shannon is 199,000,000 m^3. Lake Shannon is 11 km long, covers acres at its full pool, and has an elevation of 134 m. The Upper Baker Dam is 95 m high and impounds Baker Lake.

Glacier runoff discharge records below Sholes Glacier are compared with the observed discharge at the USGS station on the North Fork Nooksack to determine the percent of runoff generated by glaciers in 2014. The percent glacier contribution peaked on September 15 and 16 at over 80% of total stream discharge (Fig. 8.33).

From August 1 to the end of the melt season, there were 21 days with glacier runoff providing 40% or more of the total runoff. There were no days in July where this occurred, indicating that nonglacier snow melt was still important. This is an illustration of the growing seasonal importance of glacier runoff to streamflow after August 1 (Pelto, 2008). Glacier runoff peaked at 80% of streamflow on September 15. In 2014 there was minimal nonglacier snowpack remaining on September 15, and there had not been significant precipitation in the previous 14 days. In 2015 record drought

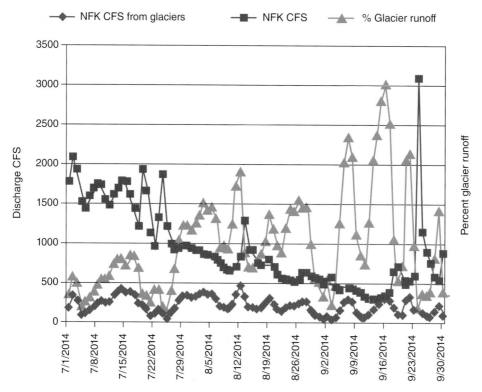

Figure 8.33 Amount of glacier runoff and percent of glacier runoff in the North Fork Nooksack River at the USGS gage, determined from the direct discharge measurements below glaciers and ablation measurements on the glaciers.

conditions led to 45 days where glacier runoff comprised more than 40% of total streamflow. It will be important to identify the impact on salmon from the ongoing salmon population surveys by NSEA (2014) in the region over the next several years.

It is evident that glaciers play a critical role in the Nooksack River in moderating fluctuations in both discharge and temperature, which reduces the stress for salmon in a system fed by glaciers (Isaak *et al.*, 2011; Mantua, Tohver, and Hamlet, 2010). The continued loss of glaciers will lead to a functional reduction in this moderating ability (Nolin *et al.*, 2010). The change in glacier area is the key component to overall glacier runoff and hence to quantifying the declines in glacier runoff (Jost *et al.*, 2012). The Skykomish and South Fork Nooksack Rivers exemplify what will occur to stream discharge as glacier area is lost in the North Fork and Middle Fork basins.

References

Abatzoglou, J.T., Rupp, D., and Mote, P.W. (2014) Seasonal climate variability and change in the Pacific Northwest of the United States. *Journal of Climate*, **27**, 2125–2142. doi: 10.1175/JCLI-D-13-00218.1

Bengston, K. (1956) Activity of the Coleman Glacier, Mt. Baker, Washington, USA., 1949–1955. *Journal of Glaciology*, **2**, 708–713.

Dery, S., Stahl, K., Moore, R. *et al.* (2009) Detection of runoff timing changes in pluvial, nival and glacial rivers of western Canada. *Water Resources Research*, **45**. doi: 1029/2008WR006975

Finn, C.A., Deszcz-Pan, M., and Bedrosian, P.A. (2012) Helicopter electromagnetic data map ice thickness at Mount Adams and Mount Baker, Washington, USA. *Journal of Glaciology*, **58** (212), 1133–1143. doi: 10.3189/2012JoG11J098

Fountain, A. and Tangborn, W. (1985) The effect of glaciers on streamflow variations. *Water Resources Research*, **21**, 579–586.

Grah, O. and Beaulieu, J. (2013) The effect of climate change on glacier ablation and baseflow support in the Nooksack River basin and implications on Pacific salmonid species protection and recovery. *Climatic Change*. doi: 10.1007/s10584-013-0747-y

Granshaw, F. and Fountain, A. (2006) Glacier change (1958–1998) in the North Cascades National Park Complex, Washington, USA. *Journal of Glaciology*, **52** (177), 251–256.

Harper, J. (1993) Glacier terminus fluctuations on Mt. Baker, Washington, USA, 1940–1980, and climate variations. *Arctic, Antarctic, and Alpine Research*, **25**, 332–340.

Harrison, A.E. (1960) *Exploring Glaciers with a Camera*, Sierra Club, San Francisco.

Heikkinen, A. (1984) Dendrochronological evidence of variation of Coleman Glacier, Mt. Baker, Washington. *Arctic, Antarctic, and Alpine Research*, **16**, 53–54.

Hubley, R. (1956) Glaciers of Washington's Cascades and Olympic Mountains: their present activity and its relation to local climatic trends. *Journal of Glaciology*, **2** (19), 669–674.

Isaak, D.J., Wollrab, S., Horan, D., and Chandler, G. (2011) Climate change effects on stream and river temperatures across the northwest U.S. from 1980–2009 and implications for salmonid fishes. *Climate Change*. doi: 10.1007/s10584-011-0326-z

Jost, G., Moore, R., Menounos, B., and Wheate, R. (2012) Quantifying the contribution of glacier runoff to streamflow in the upper Columbia River Basin, Canada. *Hydrol. Earth Syst. Sci.*, **16**, 849–860.

Key CH, Fagre DB, and Menicke RK (2002) Glacier retreat in Glacier National Park, Montana. In, R. S. Jr. Williams and Ferrigno J.G., (eds) *Satellite Image Atlas of Glaciers of the World, Glaciers of North America - Glaciers of the Western United States*. U.S. Geological Survey Professional Paper 1386-J; J365–J381.

Long, W.A. (1953) Recession of Easton and Deming Glacier. *Scientific Monthly*, **76**, 241–247.

Long, W.A. (1955) What's happening to our glaciers. *The Scientific Monthly*, **81**, 57–64.

Long, W.A. (1956) Present growth and advance of Boulder Glacier, Mt. Baker. *The Scientific Monthly*, **83**, 1–2.

Luce, C., Staab, B., Kramer, M. *et al.* (2014) Sensitivity of summer stream temperatures to climate variability in the Pacific Northwest. *Water Resources Research*, **50**, 3428–3443. doi: 10.1002/2013WR014329

Mantua, N.J., Tohver, I., and Hamlet, A. (2010) Climate change impacts on streamflow extremes and summertime stream temperature and their possible consequences for freshwater salmon habitat in Washington State. *Climate Change*, **102** (1–2), 187–223.

Mote, P., Hamlet, A., and Salathe, E. (2008) Has spring snowpack decline in the Washington Cascades? *Hydrology and Earth System Sciences*, **12**, 193–206. doi: 10.5194/hess-12-193-2008

Nolin, A.W., Phillippe, J., Jefferson, A., and Lewis, S.L. (2010) Present-day and future contributions of glacier runoff to summertime flows in a Pacific Northwest watershed: implications for water resources. *Water Resources Research*, **46** (12), W12509.

NSEA (2014) *Nooksack Salmon Enhancement Association Monitoring Program*, http://www.n-sea.org/salmon-info-1/monitoring (accessed February 2014).

Pelto, M.S. (1993) Current behavior of glaciers in the north cascades and effect on regional water supplies. *Washington Geology*, **21** (2), 3–10.

Pelto, M. (2000) Mass balance of adjacent debris-covered and clean glacier ice in the North Cascades, Washington. *IAHS Publ.*, **264**, 35–42.

Pelto, M.S. (2008) Impact of climate change on North Cascade alpine glaciers and alpine runoff. *Northwest Science*, **82** (1), 65–75.

Pelto, M.S. (2011) Skykomish River, Washington: Impact of ongoing glacier retreat on streamflow. *Hydrological Processes*, **25** (21), 3267–3371.

Pelto, M., (2015) Climate driven retreat of mount baker glaciers and changing water resources, in *SpringerBriefs in Climate Studies*, Springer International Publishing, Switzerland. doi: 10.1007/978-3-319-22605-7

Pelto, M. and Brown, C. (2012) Mass balance loss of Mount Baker, Washington glaciers 1990–2010. *Hydrological Processes*, **26** (17), 2601–2607.

Pelto, M.S. and Hedlund, C. (2001) The terminus behavior and response time of North Cascade glaciers. *Journal of Glaciology*, **47**, 497–506.

Post, A., Richardson, D., Tangborn, W.V., and Rosselot, F. (1971) Inventory of glaciers in The North Cascades, Washington. US Geological Survey Prof. Paper, 705-A.

Tennant, D.L. (1976) Instream flow regimens for fish, wildlife, recreation, and related environmental resources, in *Instream Flow Needs* vol. 2 (eds J. Osborn and C. Allman), American Fisheries Society, Western Division, Bethesda, Md., pp. 359–373.

Whatcom Farm Friends (2014) Whatcom Farm Facts. http://www.wcfarmfriends.com/#!farm-facts/citg. Accessed Feb. 2015.

9

Interior Ranges, British Columbia/Alberta

Overview

Recent inventories indicate that glacier area and volume losses are currently large and accelerating in Western Canada. Bolch, Menounos, and Wheate (2010) noted a 10% area loss for British Columbia glaciers from 1985 to 2005, and Alberta glaciers lost 25% of their area. In British Columbia losses were greatest in the interior ranges. Bolch, Menounos, and Wheate (2010) identified the losses by regions including the Southern Rockies and Southern Interior representing the region examined in this chapter and found a 15% area loss from 1985 to 2005 for both regions. Jiskoot *et al.* (2009) examined the behavior of Clemenceau Icefield and the neighboring Chaba Icefield. They observed the terminus change of 176 glaciers in the Clemenceau Icefield and adjacent Chaba Icefield and identified an average retreat of 21 m per year from the 1980s to 2001. They found that from the mid-1980s to 2001 the Clemenceau Icefield glaciers had lost 42 km^2, or 14% of total area. Tennant *et al.* (2012) examined the changes of the Rocky Mountain glaciers and found between 1919 and 2006 that glacier cover decreased by 590 km^2, 17 of 523 glaciers disappeared, and 124 glaciers fragmented into multiple ice masses. The disappearance and fragmentation are often an indication of a disequilibrium response. If a glacier does not have a consistent and persistent snow cover at the end of the melt season, it has no "income" and cannot survive (Pelto, 2010); this is the case for many glaciers in the region. Tennant and Menounos (2013) examined the changes in the Columbia Icefield during the 1919–2009 period and found a mean retreat of 1150 m and mean thinning of 49 m. The fastest rate of loss on Columbia Icefield glaciers from 1919 to 2009 was during the 2000–2009 period (Tennant and Menounos, 2013). Peyto Glacier is the only glacier in the region with a long-term annual mass balance record, which indicates a thinning of 25 m during the 1986–2015 period (Kehrl *et al.*, 2014). The impact of glacier retreat, area loss, and volume loss is of particular concern for summer water resources in the region, which have already seen a decline in glacier runoff in some basins (Moore *et al.*, 2009; Marshall *et al.*, 2011). This is a region that, as Clarke *et al.* (2015) indicate, will lose most of its glaciation by 2100.

The glaciers in the region are a crucial water resource for hydropower, irrigation, and aquatic resources (Marshall *et al.*, 2011). Jost *et al.* (2012) while examining the upper Columbia River basin noted that the basin had 5% glacier cover but that glacier melt contributed 25% and 35% of streamflow in August and September, respectively (Fig. 9.1). Marshall *et al.* (2011) used a model to quantify glacier runoff during the 2000–2007 period from the eastern slopes of the Rocky Mountains to be 0.62 ± 0.09 km^3 a^{-1}. This is equivalent to 3–4% of mean annual discharge and 7–8% of late summer (July–September) runoff in the North Saskatchewan and Bow River. Based on climate models,

Recent Climate Change Impacts on Mountain Glaciers, First Edition. Mauri Pelto.
© 2017 John Wiley & Sons, Ltd. Published 2017 by John Wiley & Sons, Ltd.

Figure 9.1 View across the Conrad Glacier in September 2015. (Courtesy of Ben Pelto.)

Marshall *et al.* (2011) indicated that glaciers on the eastern slopes will lose 80–90% of their volume by 2100. This will lead to declining glacier runoff in Alberta, from 1.1 km^3 a^{-1} in the early 2000s to 0.1 km^3 a^{-1} by the end of this century (Marshall *et al.*, 2011). Stahl and Moore (2006), in an analysis of discharge from glacierized and nonglacierized basins throughout British Columbia, determined that negative August streamflow trends indicated that the initial phase of increased runoff caused by climate warming had passed and now glacier runoff declines were part of the reason for overall runoff declines. As in the North Cascades (Pelto, 2015), Jost *et al.* (2012) concluded that the contributions are particularly important during periods of warm, dry weather following winters with low accumulation and early snowpack depletion, such as occurred in 2014 and 2015.

This chapter provides a comparison of a single Landsat image from August 12, 2015 (LC80440242015224LGN00), to a Landsat image from August 28, 1986 (LT50440241986240XXX01), except for Conrad Icefield (August 31, 1987; LT50440241987243XXX01) and Dismal Glacier (September 2, 1988; LT50440246988240XXX01) (Fig. 9.2).

9.1 Yoho Glacier

Yoho Glacier is the largest southern outflow draining the south from the Wapta Icefield in the Kootenay region of British Columbia (Fig. 9.3). It flows 6.5 km from the 3125 m to a terminus at 2200 m. The glacier terminus reach is thin, gently sloping, and uncrevassed poised for continued retreat. In an exploration of Mount Balfour in 1898, a party led by Professor Jean Habel, with the packer Ralph Edwards as a guide, was the first to visit and describe Yoho Glacier. Their descriptions of the magnificent Takakkaw Falls down river of the glacier quickly led to it becoming a frequent destination of visitors. The glacier was also accessible. Retreat up a steep slope at 2000 m made actually visiting the glacier difficult in the middle of the twentieth century. The glacier has retreated 2.1 km in the last

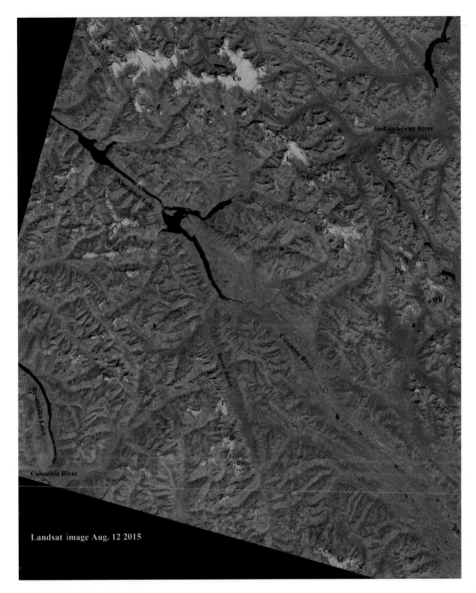

Figure 9.2 Location of glaciers examined in British Columbia and Alberta in a single Landsat image from Aug. 12, 2015 (LC80440242015224LGN00). Vowell = V, Malloy = M, Conrad = Cn, D = Deville Neve, I = Illecillewaet, G = Grand, Bv = Beaver, Dp = Des Poilus, Y = Yoho, F = Freshfield, Ci = Campbell Icefield, Co = Columbia, Wk = Waputik, L = Lyell Icefield, Mo = Mons Icefield, Ax = Apex, Cu = Cummins, T = Tusk, H = Haworth, Ss = Sir Sanford, and Di = Dismal.

century leaving a vast area of bare terrain, dotted by several small new alpine lakes. Here we examine the changes in the glacier from 1986 to 2015 with Landsat imagery.

In 1986 the glacier terminated in a broad 500 m wide glacier terminus at 2150 m (red arrow), and the glacier tongue remained wide up to the yellow arrow at 800 m. A tributary was connected to the glacier at the purple arrow, and the glacier snowline was at 2550 m. In 1998 the terminus had not retreated significantly, but had narrowed noticeably. The tributary at the purple arrow was no longer

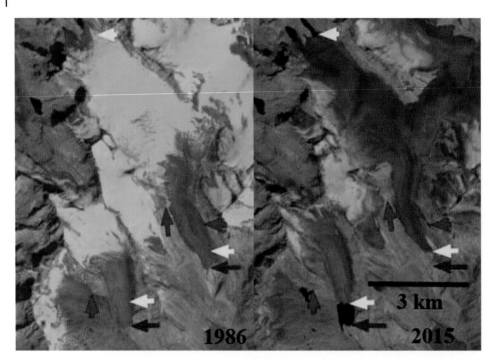

Figure 9.3 Comparison of Yoho (Y) and Des Poilus (Dp) Glaciers in Landsat images from 1986 and 2015. The 1986 terminus location is indicated by red arrow, 2015 terminus position by yellow arrow, and upglacier thinning by purple arrows.

connected, and the snowline was at 2750 m, leaving little of the glacier covered with snow, which equates to a significant mass loss. In 2013 the snowline again was high at 2700 m. In 2015 the glacier terminus has retreated 300 m since 1986 and is only 250 m wide. The width at the yellow arrow is 450 m. The width reduction is an indicator of how much the glacier has thinned. The snowline is at 2800 m in this mid-August image, clinging only to the high slopes of Mount Collie, and would still rise for several more weeks in the summer. The nearby Peyto Glacier has an annual mass balance record, indicating a thinning of 25 m during this period (Kehrl *et al.*, 2014). To be in equilibrium, a glacier typically needs more than 50% of its area to be in the accumulation zone at the end of the summer. In recent years, when the snowline exceeds 2700 m, less than 10% of the Yoho Glacier is in the accumulation zone. If the snowline is as high as it has been recently on Yoho Glacier, this indicates the lack of a significant accumulation zone, and therefore it cannot survive even the current climate. However, in both cases, the Peyto and Yoho Glaciers are rapidly losing volume but remain substantial in size and are not on the verge of disappearing in the next few decades (Fig. 9.4).

9.2 Des Poilus Glacier

Des Poilus Glacier, British Columbia, is at the southwest corner of the Wapta Icefield in Yoho National Park. The glacier flows south at a relatively low slope from the slopes of Mt. Des Poilus (Fig. 9.3). The glacier retreated 500 m, at a rate of 15–18 m per year, from 1950 to 1980. In 1986 no lake existed at the terminus (red arrow). A tributary from the west reaches the western margin of the glacier

Figure 9.4 Yoho Glacier in 2005 (no accumulation zone in sight).

(purple arrow). In 2001 a small lake had developed along the west side of the glacier, and the glacier had retreated 300 m from 1988. In 2015 a substantial lake had developed that is 640 m long and 500 m wide. The western tributary at the purple arrow has separated by over 1 km, and a small lake has developed in the deglaciated terrain. Total retreat from 1988 is 840 and 500 m since 2001. The retreat rate has increased to 25–30 m per year since 1988.

The snowline in 2013 and 2015 is very high (2650 m), which is 200 m higher than that necessary for the glacier to be in equilibrium (Fig. 9.5). The snowline, despite this being an image taken on August 12, is already at 2600 m. The issue for this glacier is that the consistent snowline is too high to support the main valley tongue of the Des Poilus Glacier, which will disappear (Pelto, 2010). The only section of the glacier that maintains snow cover is on the east flank of slopes of Mt. Des Poilus.

9.3   Waputik Icefield–Daly Glacier

Waputik Icefield is southeast of Wapta Icefield, and the main western outlet of the icefield is Daly Glacier, which drains into the Yoho River (Fig. 9.6). In 1986 Daly Glacier terminated in a proglacial lake at the red arrow. In 2001 the glacier had retreated from this lake. In 2015 the glacier terminated 1100 m from the lake. The glacier has retreated 1500 m. The purple arrows indicate areas of expanding bedrock well above the terminus region, indicating that significant thinning extends up to 2700 m. The snowline in mid-August of 2015 is above 2700 m. Across the divide, Waputik Glacier is also experiencing significant recession.

Figure 9.5 Yoho and Des Poilus Glaciers in a 2013 Landsat image.

9.4 Cummins Glacier

The Cummins Glacier is part of the Clemenceau Icefield Group in the Rocky Mountains of British Columbia (Fig. 9.7). The Cummins Glacier via the Cummins River feeds the $430\,km^2$ Kinbasket Lake, on the Columbia River. The lake is impounded by the 5946 MW Mica Dam operated by BCHydro. In 1986 Cummins Glacier had a joint terminus with the main southeast flowing branch and the west flowing branch terminating at the red arrow. The glacier also had a substantial connection (purple arrow) with Tusk Glacier that flows east terminating northeast of Tusk Peak. There are other connections with other high-elevation accumulation areas (purple arrows). In 2013 and 2014 Cummins Glacier had less than 20% retained snow cover by the end of the melt season. Typically 50–65% of a glacier must be covered with snow at the end of the summer season to be in equilibrium. In 2015 conditions were even worse with no retained snow cover; in fact, there are only minor patches of retained firn from previous years. The lack of a persistent accumulation zone indicates a glacier that cannot survive the climate conditions (Pelto, 2010). In 2015 a proglacial lake had formed at the terminus that is 500 m long, representing the retreat during the 30-year period. The west flowing portion of the Cummins has detached from the larger branch. The connection to Tusk Glacier is nearly severed

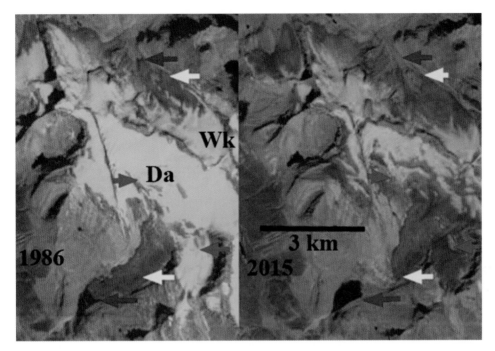

Figure 9.6 Comparison of Daly (Da) and Waputik (Wk) Glaciers in Landsat images from 1986 and 2015. The 1986 terminus location is indicated by red arrow, 2015 terminus position by yellow arrow, and upglacier thinning by purple arrows.

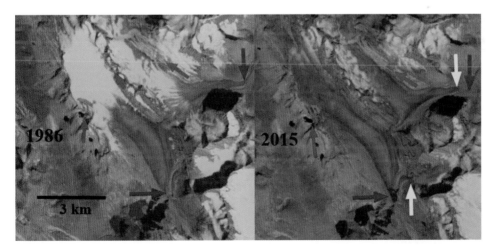

Figure 9.7 Comparison of Cummins (C) and Tusk (T) Glaciers in Landsat images from 1986 and 2015. The 1986 terminus location is indicated by red arrow, 2015 terminus position by yellow arrow, and upglacier thinning by purple arrows.

and, in terms of flow, is effectively ended. Retreat of the margin higher on the glacier is also evident at each purple arrow. Tusk Glacier is no longer connected to Duplicate Glacier and has retreated to the north side of Tusk Peak. The dominant change in Cummins Glacier has been thinning; it should now be poised for a more rapid retreat.

The result for Kinbasket Lake of the loss of the collective large area is a reduction in summer glacier runoff. The annual runoff which will be dominated by annual precipitation would not change just because of the glacier loss as noted in cases like the Skykomish Basin (Pelto, 2011) and on Bridge River (Stahl *et al.* 2008).

9.5 Apex Glacier

Apex Glacier drains into the Wood River and then into the 430 km^2 Kinbasket Lake, on the Columbia River. The lake is impounded by the 5946 MW Mica Dam operated by BCHydro. In a recent paper Jiskoot *et al.* (2009) examined the behavior of glacier in this and the neighboring Chaba Icefield. They found that from the mid-1980s to 2001 the Clemenceau Icefield glaciers had lost 14% of their area. During this same period, terminus retreat of the icefield glaciers averaged 21 m a^{-1}. In 1986 Apex Glacier ended at the red arrow with no lake at the terminus (Fig. 9.8). In 2001, in Google Earth images, a lake had formed that is 0.5 km long, and in 2015, the lake is 0.7 km long and the total glacier retreat is 800 m. The reduced rate of retreat since 2010 may result from the lake depth being reduced at the current terminus; however, the glacier will continue to retreat as upglacier thinning is significant. Another interesting aspect is that this glacier is fed by a northern and a southern accumulation zone. The remaining accumulation zone is quite small for the northern accumulation zone with another month of melting to go. In 2013 the glacier lost all but 10% of its snow cover with the snowline at 2700 m. This represents a large negative mass balance that will reinforce retreat. In 2015 the snowline is also above 2700 m.

9.6 Shackleton Glacier

Shackleton Glacier drains south from the Clemenceau Icefield into the Kinbasket River. In 1986 the glacier terminated at 1200 m, the lowest elevation terminus of the glaciers examined in this chapter (Fig. 9.8). The glacier descends the southern flank of Shackleton Mountain and has a steep narrow icefall from 1900 to 1400 m. In 2015 the terminus had retreated 850 m and terminated at 1400 m. The upglacier thinning is extensive (purple arrows), indicating a continued rapid retreat. The snowline on this glacier in 2015 is at 2400 m lower than the surrounding glaciers, despite the southern orientation. This is also the case in 2013, suggesting the glacier has an ideal orientation to promote high accumulation. Just southeast of the main glacier is an unnamed glacier (Se) that ends in a basin at 1800 m in 1986. In 2015 the glacier had lost most of the terminus lobe that occupied the basin. The main tributary has also detached at the purple arrow.

9.7 Columbia Glacier

The Columbia Glacier drains west from Columbia Icefield into British Columbia. The glacier was 8.5 km long in 1964, 9.5 km long in 1980, and 6.2 km long in 2015. The glacier drops rapidly from the plateau area over a major icefall from 2400 to 1950 m, which created a series of ogives during the 1960–1990 period. Ogives are annual wave bulges that form at the base of an icefall due

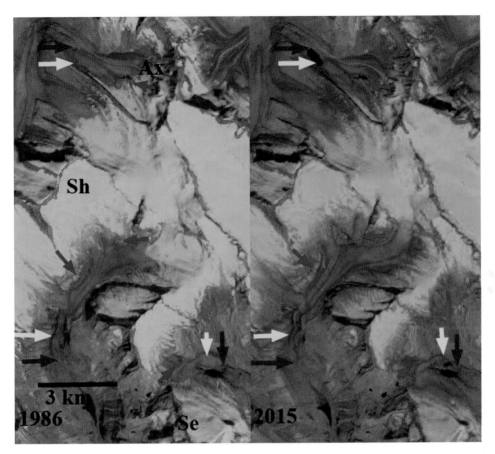

Figure 9.8 Comparison of Apex (Ax) and Shackleton (Sh) Glaciers, and an unnamed glacier (Se) in Landsat images from 1986 and 2015. The 1986 terminus location is indicated by red arrow, 2015 terminus position by yellow arrow, and upglacier thinning by purple arrows.

to differential seasonal flow velocity. Ommanney (2002) noted that the glacier advanced over 1 km from 1966 to 1980 and the glacier completely filled the large proglacial lake that now exists. In 1986 retreat had again opened the lake (Fig. 9.9). Tennant and Menounos (2013) examined the changes in the Columbia Icefield during the 1919–2009 period and found a mean retreat of 1150 m and mean thinning of 49 m.

In 1986 the lake is 1000 m long. In 2015 the lake is 4000 m long, indicating a 3000 m retreat from 1986 to 2015. The rate of retreat has been less since 2004 (300 m), as the glacier approaches the upper limit of the lake basin. A 2004 Google Earth image indicates a step in elevation that is 500 m from the terminus. Glacier elevation lags the basal elevation change; hence, the end of the lake is between 500 and 1000 m from the 2004 terminus. When the glacier terminus retreats to this step, the lake will no longer enhance retreat via calving and retreat rates will diminish. A further change is noted in the absence of ogives at the base of the icefall (Fig. 9.10). As the icefall has narrowed and slowed, the result has been a cessation of this process. The purple arrow indicates a tributary that joined the glacier below the icefall in 1986 that now has a separate terminus. The current terminus is still active with crevassing near the active front.

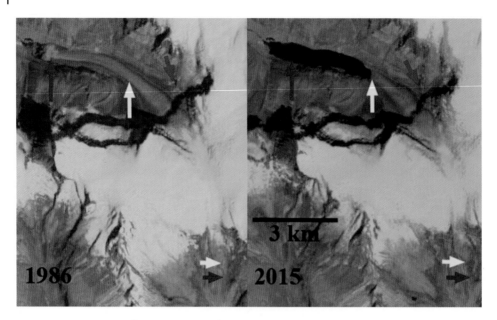

Figure 9.9 Comparison of Columbia Glacier from 1986 and 2015. The 1986 terminus location is indicated by red arrow, 2015 terminus position by yellow arrow, and upglacier thinning by purple arrows.

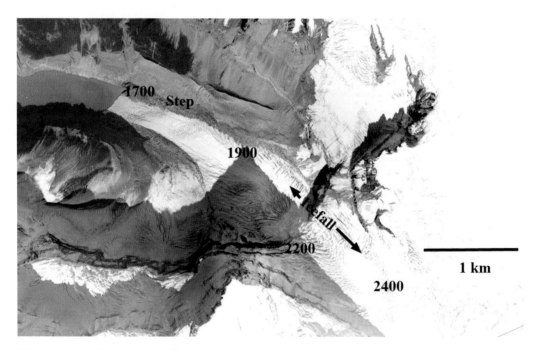

Figure 9.10 Google Earth image of Columbia Glacier from 2004 indicating elevations and the position of the icefall. (Google Earth.)

9.8 Freshfield Glacier

Freshfield Glacier drains north into Freshfield Lake one of the headwaters of Saskatchewan River. The Freshfield Glacier in the Canadian Rockies retreat over the last 5 years has exposed a new glacier lake. Today the glacier is 10.1 km long beginning at 3070 m and ending at 1970 m near the shore of the less than 5-year-old lake. This glacier during the Little Ice Age stretched 14.3 km, one of the longest in the entire range. In 1964 the glacier had retreated 1900 m exposing Freshfield Lake. From 1964 to 1994, the glacier retreated up this lake basin losing another 1500 m of length. In 1986 the lake was 1500 m long (Fig. 9.11). In the mid-1990s the glacier no longer reached the shores of Freshfield Lake. In 2005 a Google Earth image indicates the glacier had retreated 400 m from the lake and a New Lake had developed that is 370 m across. In 2005, above the terminus, there were two concentric depressions that typically indicate a depression beneath the glacier that would tend to at least seasonally fill with water (purple arrows; Fig. 9.12). Such a depression cannot form except with a stagnating and rapidly retreating glacier tongue. In 2015 the glacier no longer reached even the New Lake. The glacier has retreated 1600 m since 1986. The rate of retreat has remained substantial despite the lack of calving since the late 1990s.

The 2013 Landsat image indicates that the snowline is high at 2550–2600 m (Fig. 9.13). The snowline was even higher in 2014 and 2015, above 2600 m. The sustained high elevation of the snowline will lead to continued retreat of Freshfield Glacier. The purple arrows indicate a number of locations where bedrock has expanded as the glacier has contracted.

9.9 Lyell Icefield – Mons Icefield

Lyell Icefield straddles the Alberta – British Columbia border. The Lyell Icefield's western terminus drains into the Valenciennes River and then the Columbia River. The Lyell Icefield eastern terminus and the Mons Icefield drain into the Glacier River and then into the Saskatchewan River. In 1986 the western outlet of Lyell Icefield ended at an elevation of 1850 m (Fig. 9.14). In 2015 the glacier had retreated 300 m terminating at an elevation of 2000 m. The eastern outlet of Lyell Icefield terminated in a basin at 1540 m, a short distance above the lake. In 2015 the terminus lobe in the basin had melted, and the glacier retreated 500 m since 1986, which terminated at an elevation of 1650 m. The upper section of the Lyell Icefield has a number of areas of expanded bedrock high on the glacier indicating thinning. The snowline is at 2650 m in 2015 – more of the Lyell Icefield is covered with snow than other glaciers in the region. The Mons Icefield in 1986 terminated at the top of a prominent gully at an elevation of 1850 m. In 2015 the glacier had retreated 750 m and terminated at 2000 m. A section of the glacier extending southeast from the main glacier tongue has detached from the main glacier (purple arrow).

9.10 Haworth Glacier

Haworth Glacier in the northern Selkirk Mountains of British Columbia drains into Palmer Creek, which flows into Kinbasket Lake, and then the Columbia River. This glacier is often visited by climbers as the Canadian Alpine Club has a summer base camp near the terminus of the glacier. The glacier has a low slope and limited crevassing that make it a good training ground for climbing. Menounos *et al.*

Figure 9.11 Comparison of Freshfield Glacier in Landsat images from 1986 and 2015. The 1986 terminus location is indicated by red arrow, 2015 terminus position by yellow arrow, and upglacier thinning by purple arrows.

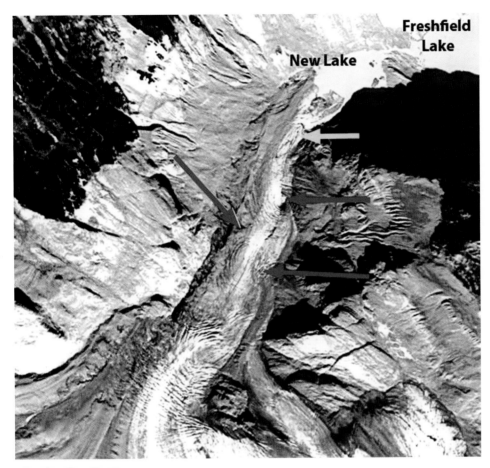

Figure 9.12 Google Earth image from 2005, indicating circular depressions where lakes will form with continued retreat. (Google Earth.)

(2008) noted an advance of this glacier overrunning a stump that has since been exposed by retreat 3800 years before present, similar in timing to many glaciers in the region. The stump remained buried until recent exposure.

Here we examine a series of Landsat images from 1986 to 2015 to identify the retreat and to forecast whether the glacier can survive even current climate conditions (Fig. 9.15). In 1986 the glacier ended near the far end of the basin where a lake has since developed (red arrow). Snow covers 30% of the glacier in the late summer of 1986, while 55–65% is necessary for glacier equilibrium. In 1998 the glacier had retreated 550 m since 1986, and the glacier was 15% snow covered. In 2009 the glacier was 20% snow covered. In 2015 the glacier terminated at the yellow arrow, indicating a retreat of 1000 m since 1986, 37 m per year. The glacier was approximately 10% snow covered in 2013 (Fig. 9.16) and less than 10% snow covered in 2015. The glacier has less than 10% of its area above 2500 m, and this is where most of the retained snow cover is. The percent snow cover each year is much less than the 55% minimum needed for a minimum balance; the images are also not precisely at the end of the melt season. If a glacier does not have a consistent and persistent snow cover at the end of the melt season,

Figure 9.13 A 2013 Landsat image of Freshfield Glacier with the snowline indicated by blue dots. This is a pan-sharpened Landsat image created by Ben Pelto (UNBC). (Courtesy of Ben Pelto.)

it has no "income" and cannot survive (Pelto, 2010). This glacier has managed to retain a very small area of snow cover, but given the ongoing thinning and the lack of avalanche accumulation on this glacier, it is unlikely to be enough to save this glacier. Bolch, Menounos, and Wheate (2010) noted a 10% area loss for British Columbia glaciers from 1985 to 2005; Haworth Glacier is above this average. Tennant and Menounos (2013) noted that the fastest rate of loss on Columbia Icefield glaciers from 1919 to 2009 was during the 2000–2009 period.

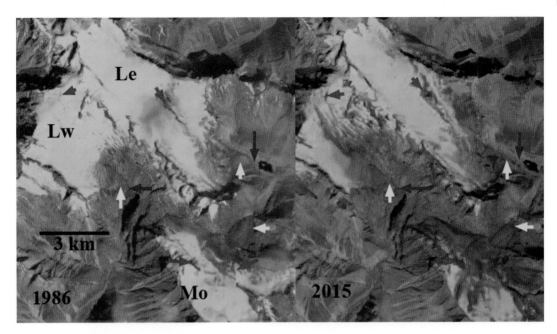

Figure 9.14 Comparison of Lyell Icefield west (Lw), Lyell Icefield east (Le), and Mons Icefield (Mo) in Landsat images from 1986 and 2015. The 1986 terminus location is indicated by red arrow, 2015 terminus position by yellow arrow, and upglacier thinning by purple arrows.

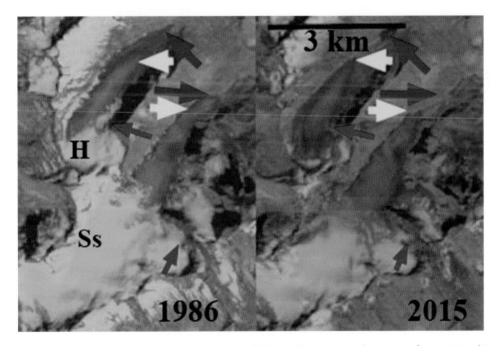

Figure 9.15 Comparison of Haworth (H) and Sir Sanford (Ss) Glaciers in Landsat images from 1986 and 2015. The 1986 terminus location is indicated by red arrow, 2015 terminus position by yellow arrow, and upglacier thinning by purple arrows.

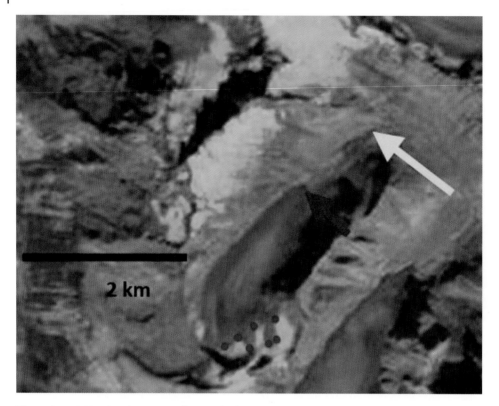

Figure 9.16 A 2013 Landsat image of Haworth Glacier, indicating terminus and snowline.

9.11 Sir Sandford Glacier

This glacier flows north from the slopes of Mount Sir Sandford and drains west into Kinbasket Lake. The glacier has a higher elevation accumulation zone than the adjacent Haworth Glacier, with a considerable area above 2500 m. From 1986 to 2015, the glacier terminus retreated 600–700 m. In years 1986, 1998, 2009, 2013, and 2015, snow-covered area remained above 25% of the glacier area, indicating a persistent accumulation zone. This glacier can survive current climate. The purple arrow indicates areas of thinning higher on the glacier.

9.12 Dismal Glacier

Dismal Glacier flows north from Mount Durand in the Selkirk Range of British Columbia. It drains from 2500 to 1950 m, and its runoff flows into Downie Creek that is a tributary to the Columbia River and Revelstoke Lake. This lake is impounded by BC Hydro's Revelstoke Dam which is a 2480 MW facility. Here we examine Landsat images from 1988 and 2015 to identify changes in this glacier (Fig. 9.17). The glacier snowline in the mid-August image of 2015 is at 2400 m just above a substantial icefall. The glacier had retreated 640 m from 1988 to 2015. The eastern extension at 2200–2300 m of

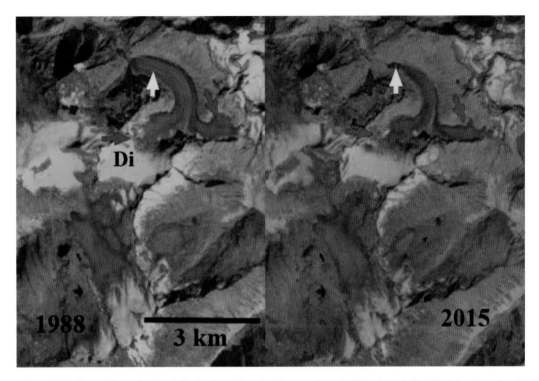

Figure 9.17 Comparison of Dismal Glacier (Di) in Landsat images from 1988 and 2015. The 1988 terminus location is indicated by red arrow, 2015 terminus position by yellow arrow, and upglacier thinning by purple arrows.

the glacier noted by a purple arrow has lost considerable area, indicating thinning even well above the terminus elevation. Note the thinning of this section of the glacier in 2015; after it joined the main glacier, it is separated by a medial moraine. The terminus in the 2009 Google Earth image has a low slope and is uncrevassed. This indicates that the terminus reach is relatively inactive, but does not appear stagnant. Bolch, Menounos, and Wheate (2010) observed a 15% area loss from 1985 to 2005 in this region. The snowline has been above the icefall at 2400+ m in 2013, 2014, and 2015, indicative of negative mass balance that will lead to continued retreat. The glacier's name is not due to its future prospects, but its future prospects are indeed dismal.

9.13 Illecillewaet Icefield

Illecillewaet Icefield is just south of Rogers Pass. Illecillewaet Glacier is the main northern outlet of the icefield and drains into a river of the same name, which is a tributary of the Columbia River. It spills down from the high plateau above 2500 m. The high plateau stretches 5 km north to south and also feeds Geikie Glacier, the main southern outlet of the icefield. The glacier retreated rapidly from 1877 to 1950 and then did not retreat until after 1980. In 1986 the glacier terminated at the red arrow at 1900 m. In 2015 the terminus had retreated upslope to an elevation of 2000 m and a distance of 250 m (Fig. 9.18).

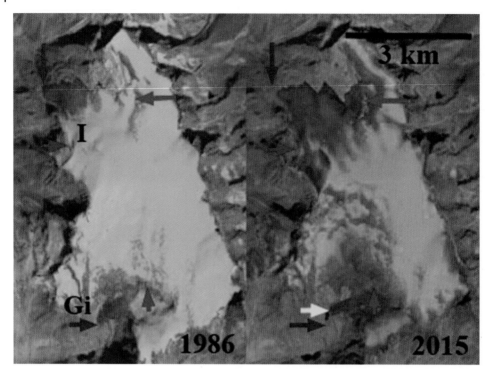

Figure 9.18 Comparison of Illecillewaet (I) and Geikie (Gi) Glaciers in Landsat images from 1986 and 2015. The 1986 terminus location is indicated by red arrow, 2015 terminus position by yellow arrow, and upglacier thinning by purple arrows.

Geikie Glacier followed the same pattern with a noticeable expansion from 1950 to 1980 (Ommanney, 2002). In 1986 the terminus, which drains into the Incomappleux River, a tributary of the Columbia River, was at 1825 m. In 2015 the glacier had retreated to over 1900 m in elevation a distance of 330 m. A small narrow proglacial lake has developed at the terminus. The lake is too small to impact retreat rate.

The purple arrows indicate two areas of thinning at approximately 2400 m. Even in the warm summers of 2015, the snowline remained close to 2500 m, and the glacier had a significant retained snow cover. This explains why the retreat has been slower; however, the snowline has still been too high to maintain glacier size.

9.14 Deville Icefield

Deville Icefield is just south of Illecillewaet Icefield; the main northern outlet is Deville Glacier, which drains into the Incomappleux River. The glacier has a substantial icefall that extends from 2200 to 1900 m (orange arrow). In 1986 the glacier terminus at the red arrow was fed by a glacier that is 400 m wide and extends for 1 km north from the base of the icefall (Fig. 9.19). In 2015 the terminus tongue below the icefall had been reduced to a width less than 200 m and a length of 600 m. The reduction in area has been more pronounced than the 400 m retreat of the terminus. A small lake has developed at the terminus.

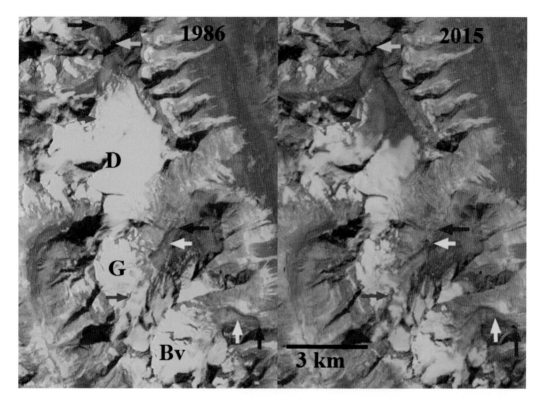

Figure 9.19 Comparison of Deville (D), Grand (G), and Beaver (Bv) Glaciers in Landsat images from 1986 and 2015. The 1986 terminus location is indicated by red arrow, 2015 terminus position by yellow arrow, and upglacier thinning by purple arrows.

Grand Glacier is in the southeast flank of Grand Mountain and flows east into the Beaver River watershed. The glacier retreated 600 m from 1986 to 2015. The ablation zone of the glacier has narrowed and expanded upglacier. The snowline in 2015 in mid-August was at 2400 m, with an area of expanding bedrock (purple arrow) evident. Beaver Glacier is the next significant glacier south of Grand Glacier and is at the headwaters of the Beaver River. A low slope terminus tongue extended from 1750 to 1600 m in 1986. In 2015 this entire tongue had been lost, and the glacier had retreated 750 m and terminated at 1750 m. The snowline is at 2350–2400 m.

9.15 Conrad Icefield

Conrad Icefield is at the northern edge of the Bugaboos in the Purcell Range of the Selkirk Mountains in southwest British Columbia. The icefield feeds several terminus tongues, primarily the Conrad Glacier and Malloy Glacier; both of these have two arms. In the case of Conrad Glacier, the two arms still join above the terminus, while for Malloy Glacier there are now separate termini.

In 1987 Conrad Glacier terminus extended to the red arrow and no proglacial lake is present (Fig. 9.20). Both the south and west termini of Malloy Glacier reach the shore of a small lake. In 2005 both termini had retreated from the lake shore (Fig. 9.21). A close-up view of the Malloy Glacier terminus (red arrows) from 2005 indicates that the southern terminus is quite stagnant below the

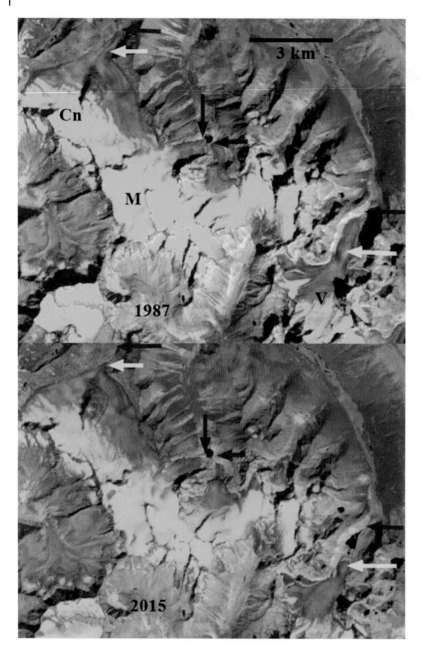

Figure 9.20 Comparison of Conrad (Cn), Malloy, and Vowell (V) Glaciers in Landsat images from 1987 and 2015. The 1987 terminus location is indicated by red arrow, 2015 terminus position by yellow arrow, and upglacier thinning by purple arrows.

Figure 9.21 Google Earth image of Malloy Glacier terminus. (Google Earth.)

icefall and retreat is continuing. The west terminus is quite narrow and ending on a steep slope, with a buttress paralleling the north side of the terminus ending at the orange arrow.

The Conrad Glacier terminus is stagnant beyond the knob at Point A and is only 200 m wide from Point A to Point B (Fig. 9.22). The proglacial lake beyond the Conrad Glacier terminus is now 400 m long and the terminus has just retreated upvalley of the lake. Conrad Glacier has retreated a significant distance from the proglacial lake and ended on a line between the knob at Points A and B. In 1987 the Conrad Glacier extended 1700 m after the two arms joined. In 2015 this distance is ~700 m (Fig. 9.23). The upstream location of the joining has changed slightly, indicating a retreat of 950 m from 1987 to 2015. Another measure is the distance from where the two arms of Conrad Glacier join the terminus. In 2015 Malloy Glacier western terminus was at the top of the buttress, with orange arrow ending 300 m from the lake (yellow arrow). The southern terminus has pulled back 250–300 m from the lake. Malloy Glacier had undergone a 300 m retreat from 1986 to 2015.

Examination of the margins of both glaciers above the icefalls 1 km above the terminus in 2005 indicates thinning and downwasting (red arrows), suggesting reduced flow that will drive continued retreat, but not as fast as Vowell Glacier, also in the Bugaboos.

9.16   Vowell Glacier

Vowell Glacier is the largest glacier of the Bugaboo's or maybe was. The glacier drains to the north into Vowell Creek and had retreated quite rapidly from 1998 to 2013, creating a new lake and then retreating from that lake. In 1986 a Landsat imagery showed the glacier to be 5.7 km long ending at 2040 m with no sign of a lake. In 2005 a Google Earth imagery indicates an 850 m retreat since

Figure 9.22 Google Earth image of Conrad Glacier terminus. (Google Earth.)

Figure 9.23 Conrad Glacier terminus in September 2015 (Ben Pelto, UNBC). Note the thin nature of the glacier below the junction and the lack of crevasses. (Courtesy of Ben Pelto.)

1986 and the formation of a proglacial lake with numerous icebergs and residual glacier pieces. The lower 500 m of the glacier is stagnant, uncrevassed, and thin in 2005. In 2015 the glacier no longer terminated in a lake and has retreated 1500 m since 1986.

The glacier is flowing with some vigor through the icefall that extends from 2400 to 2200 m. The lower 400 m of the glacier still appears stagnant and poised for retreat. Worse is the fact that the image from August 22, 2013, with a month of melting to go, shows the snowline at 2600 m. In 2015 the snowline on August 12 was again above 2600 m.

References

Bolch, T., Menounos, B., and Wheate, R. (2010) Landsat-based inventory of glaciers in western Canada, 1985–2005. *Remote Sensing of Environment*, **114** (1), 127–137. doi: 10.1016/j.rse.2009.08.015

Clarke, G., Jarosch, A., Anslow, F. *et al.* (2015) Projected deglaciation of western Canada in the 21st century. *Nature Geoscience.* doi: 10.1038/ngeo2407

Jiskoot, H., Curran, C.J., Tessler, D.L., and Shenton, L.R. (2009) Changes in Clemenceau Icefield and Chaba Group glaciers, Canada, related to hypsometry, tributary detachment, length, slope and area-aspect relations. *Annals of Glaciology*, **50**, 133–143.

Jost, G., Moore, R.D., Menounos, B., and Wheate, R. (2012) Quantifying the contribution of glacier runoff to streamflow in the upper Columbia River Basin. *Canada Hydrology and Earth System Sciences*, **16**, 849–60.

Kehrl, L., Hawley, R., Osterberg, D. *et al.* (2014) Volume loss from lower Peyto Glacier, Alberta, Canada, between 1966 and 2010. *Journal of Glaciology*, **60** (219), 2014. doi: 10.3189/2014JoG13J039

Marshall, S.J., White, E.C., Demuth, M.N. *et al.* (2011) Glacier water resources on the eastern slopes of the Canadian Rocky Mountains. *Canadian Water Resources Journal*, **36**, 109–134. doi: 10.4296/cwrj3602823

Menounos, B., Clague, J., Osborn, G. *et al.* (2008) Western Canadian glaciers advance in concert with climate change circa 4.2 ka. *Geophysical Research Letters*, **35**, L07501. doi: 10.1029/2008GL033172

Moore, R.D., Fleming, S.W., Menounos, B. *et al.* (2009) Glacier change in western North America: influences on hydrology, geomorphic hazards and water quality. *Hydrological Processes*, **23**, 42–61.

Ommanney, C.S.L. (2002) Glaciers of the Canadian Rockies J199–289, in *Satellite Image Atlas of Glaciers of the World: North America, United States Geological Survey Professional paper 1386-J—1* (eds R.S. Williams Jr., and J.G. Ferrigno), US Government Printing Office, Washington, DC.

Pelto, M. (2010) Forecasting temperate alpine glacier survival from accumulation zone observations. *The Cyrosphere*, **3**, 323–350.

Pelto, M. (2011) Skykomish River, Washington: impact of ongoing glacier retreat on streamflow. *Hydrologic Processes*, **25** (21), 3356–3363. doi: 10.1002/hyp.8218

Pelto, M. (2015) *Climate Driven Retreat of Mount Baker Glaciers and Changing Water Resources*, Springer Briefs in Climate Series, London, UK.

Stahl, K. and Moore, D. (2006) Influence of watershed glacier coverage on summer streamflow in British Columbia, Canada. *Water Resources Research*, **42** (W6), W06201. doi: 10.1029/2006WR005022

Stahl, K., Moore, R.D., Shea, J.M. *et al.* (2008) Coupled modelling of glacier and streamflow response to future climate scenarios. *Water Resources Research*, **44**, W02422. doi: 10.1029/2007WR005956

Tennant, C. and Menounos, B. (2013) Glacier change of the Columbia Icefield, Canadian Rocky Mountains, 1919–2009. *Journal of Glaciology*, **59** (216), 671–686. doi: 10.3189/2013JoG12J135

Tennant, C., Menounos, B., Wheate, R., and Clague, J.J. (2012) Area change of glaciers in the Canadian Rocky Mountains, 1919 to 2006. *The Cryosphere*, **6** (6), 1541–1552. doi: 10.5194/tc-6-1541-2012

10

Himalaya

Overview

In the Himalayan Range, stretching from the Karakoram Range in northwest India east–southeast to the border region of Bhutan and China, detailed glacier mapping inventories, from the Global Land Ice Measurements from Space (GLIMS), International Centre for Integrated Mountain Development (ICIMOD), Indian Space Research Organization (ISRO), and China National Committee for the International Association of Cryospheric Sciences (IACS), of over 10,000 glaciers have indicated increased strong thinning and area loss since 1990 throughout the Himalayan Range. The inventories rely on repeat imagery from ASTER, Corona, Landsat, IKONOS, and SPOT. It is simply not possible to make observations on this number of glaciers in the field. ICIMOD (2014) has provided an application for viewing individual glacier boundaries in Nepal.

In Garhwal Himalaya, India, of the 58 glaciers examined from 1990 to 2006, the area loss was 6% (Bhambri *et al.*, 2011). They also noted that the number of glaciers increased from 69 (1968) to 75 (2006) due to the fragmentation of ice bodies. Examination of 466 glaciers in the Chenab, Parbati, and Baspa Basins, India, illustrated a 21% decline in glacier area from 1962 to 2004 (Kulkarni *et al.*, 2007). Glacier fragmentation was also observed in this study, which for some fragments represents a loss of the accumulation area, which means the glacier will not survive (Pelto, 2010). The India glacier inventory (Kulkarni and Karyakarte, 2014) identified glacier area losses and frontal change on 2190 glaciers and found an area loss rate of 3.3% per decade and 76% of glaciers retreating. Kulkarni and Karyakarte (2014) report on Indian Himalayan glaciers that 79 of 80 with terminus change records have been receding.

In the Nepal Himalayas, area loss of 3808 glaciers from 1963 to 2009 is nearly 20% (Bajracharya and Shrestha, 2011). The Langtang subbasin is a small northeast–southwest elongated basin, tributary of Trishuli River north of Kathmandu and bordered with China to the north. The basin contained 192 km^2 of glacier area in 1977, 171 km^2 in 1988, 152 km^2 in 2000, and 142 km^2 in 2009. In 32 years from 1977 to 2009, the glacier area declined by 26% (Bajracharya and Shrestha, 2011). In the Khumbu region, Nepal, volume losses increased from an average of 320 mm a^{-1} during 1962–2002 to 790 mm a^{-1} during 2002–2007, including area losses at the highest elevation on the glaciers (Bolch, Pieczonka, and Benn, 2011). The Dudh Koshi Basin is the largest glacierized basin in Nepal. It has 278 glaciers of which 40, amounting to 70% of the area, are valley type. Almost all the glaciers are retreating at rates of 10–59 m a^{-1} and the rate has accelerated after 2001 (Bajracharya and Mool, 2009). Baidya (2014) completed an inventory of Nepal glaciers and found a 21% decline in area from the 1970s to 2007/2008.

Recent Climate Change Impacts on Mountain Glaciers, First Edition. Mauri Pelto.
© 2017 John Wiley & Sons, Ltd. Published 2017 by John Wiley & Sons, Ltd.

An inventory of 308 glaciers in the Nam Co Basin, Tibet, noted an increased loss of area of 6% for the 2001–2009 period (Bolch *et al.*, 2010). Zhou *et al.* (2009) looking at the Nianchu River Basin, southern Tibet, found a 5% area loss during 1990–2005. In the Pumqu Basin, Tibet, an inventory of 999 glacier from 1974–1983 to 2001 indicated a loss of 9% of the glacier area and a 10% disappearance of the glaciers (Jin *et al.*, 2005). The high elevation loss is also noted in Tibet on Naimona'nyi Glacier which has not retained accumulation even at 6000 m. This indicates a lack of high-altitude snow–ice gain (Kehrwald *et al.*, 2008).

A new means of assessing glacier volume is GRACE, which cannot look at specific changes of individual glaciers or watersheds. In the high mountains of Central Asia, GRACE imagery found mass losses of $-264 \, \text{mm a}^{-1}$ for the 2003–2009 period (Matsuo and Heki, 2010). This result is in relative agreement with the other satellite image assessments, but is at odds with the recent global assessment from GRACE, which estimated Himalayan glacier losses at 10% of that found in the afore-mentioned examples for volume loss for the 2003–2010 period (Jacob *et al.*, 2012). At this point the detailed glacier-by-glacier inventories of thousands of glaciers are better validated and illustrate the widespread significant loss in glacier area and volume, though not all glaciers are retreating. Yao *et al.* (2012) in an examination of Tibetan glaciers observed substantial losses of 7090 glaciers. Bolch *et al.* (2012) in a report on the "State and Fate of Himalayan Glaciers" noted that most Himalayan glaciers are losing mass and retreating at rates similar to the rest of the globe.

The glacier response to climate change is important because of the impact on water resources that can be critical to both hydropower and agriculture in the region (Pelto, 2011; WWF, 2005). The importance of glacier meltwater contribution varies by basin and timing of the primary accumulation season. The role declines in importance – from extremely important in the Indus Basin, important for the Brahmaputra Basin, and relatively unimportant in the Ganges, Yangtze, and Yellow Rivers (Immerzeel, Beek, and Bierkens, 2010). By region, meltwater contributes 30% to the total water flow in the eastern Himalayas, 50% in the central and western Himalayas, and 80% in Karakoram (Xu, Shrestha, and Eriksson, 2009).

Debris cover is a key variable affecting the rate of glacier response; it slows surface melt and retreat in the ablation zone (Bajracharya, Mool, and Shrestha, 2007; Basnett, Kulkarni, and Bolch, 2013; Bajracharya, Maharjan, and Shresta, 2014). The other key factor is the development of a proglacial lake at the terminus that accelerates melting (Basnett, Kulkarni, and Bolch, 2013; Bajracharya, Maharjan, and Shresta, 2014).

In Li *et al.* (2011), it is noted that at increasing temperature, especially at altitude, the fronts of 32 glaciers in Sikkim have retreated, mass losses of 10 glaciers have been considerable, glacial lakes in 6 regions have expanded, and meltwater discharge of 4 basins has also increased. Neckel *et al.* (2014) examined changes in surface elevation of the glaciers and found that this region lost $440 \, \text{mm a}^{-1}$ from 2003 to 2009.

In this chapter I cannot examine glaciers from all portions of the Himalayan Range. The focus is on single region that straddles the Nepal–Sikkim–China border (Fig. 10.1). This region provides a transect from the drier/colder climate in China to the wetter/warmer climate of Nepal and Sikkim. All the glaciers are summer accumulation-type glaciers. This means that the glacier receives most (~80%) of its snowfall during the summer monsoon (Ageta and Higuchi, 1984). This is also the period when ablation – low on the glacier – is highest. Following the summer monsoon which ends in early September, there is a transition period with some colder storm events where the snowline drops. Then, from November to February is the dry winter monsoon with limited precipitation. Thus, a difference compared to most glaciers, as winter proceeds, often the lower glacier remains snow-free. The premonsoon season from March to May features increasing precipitation, temperature, and rising snowlines.

Figure 10.1 Overview of Himalayan region examined in a 2015 Landsat image. Z = Zemu, C = Changsang, E = East Langpo, SL = South Lhonak, ML = Middle Lhonak, NL = North Lhonak, Ka = Kaer, Lb = Longbashaba, Zh = Zhizhai, J = Jimi, Y = Yindapu, G = Gelhaipuco, Q = Qangzonkco, N = Nobuk, Np = Nangama Pokhari, and K = Kanchenjunga.

Of the 16 glaciers examined, 15 retreated, and 13 terminated in glacier moraine-dammed lakes, which have been expanding due to retreat. Zemu Glacier, the largest glacier, is the one glacier that is not retreating. Zemu Glacier and Kanchenjunga Glacier, the second largest, start at the highest elevations and have the greatest debris cover. In the Tamor River Basin, Nepal, construction is underway on the Tamor–Mewa 110 MW hydropower project, with two more plants under development. In the Teesta River Basin, Sikkim, there are several hydropower plants already operating and many proposed. Glacier retreat has to date led to the formation of many additional glacier moraine-dammed lakes. The moraines are not made of stable material; hence the outlets can fail, leading to glacier lake outburst floods that have been a focus of much research by ICIMOD (2011).

10.1 Middle Lhonak Glacier

Middle Lhonak Glacier is an unnamed glacier between North Lhonak Glacier and South Lhonak Glacier flowing west from Janak Peak draining into the Teesta River, which has several existing and many proposed hydropower projects, mostly run-of-river with minor dams. There is a substantial icefall that begins at 5800–5900 m, and the glacier terminates in a proglacial lake (Fig. 10.2).

In 1991 Middle Lhonak Glacier ended in a proglacial lake at the red arrow, with the terminus being fed by two significant arms. In 2013 the green arrow indicated where the two arms of the glacier previously joined (Fig. 10.3). In 2015 the southern arm of the glacier no longer connected to the main glacier. Significant thinning has occurred upglacier at the purple arrows. Retreat from 1991 to 2015 is 1100 m. Above the icefall, the glacier is almost always covered with snow, but in the icefall the glacier often remains snow-free for much of the year at around 5700 m, as indicated by recent Landsat images from November to January. In 2013 a series of images indicated the snowline in a period from October 12, 2013, to December 21, 2013. On October 21 the snowline is at the last bend above the terminus at 5650 m. On November 21 the snowline has shifted a little. By December 1, the

Figure 10.2 Changes in South Lhonak (SL), Middle Lhonak (ML), and North Lhonak (NL) Glaciers from 1991 to 2015. Red arrow indicates the 1991 terminus, yellow arrow the 2015 terminus, and purple arrow areas of thinning.

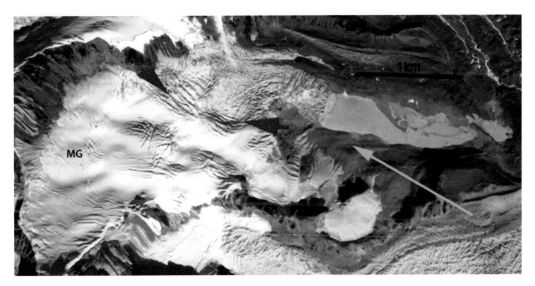

Figure 10.3 Google Earth Image in 2013. Green arrow indicates detachment o f southern arm of Middle Lhonak Glacier. Pink dots indicate moraine-dammed lake. Red arrows indicate the top of icefall and typical snowline in December. There is considerable glacier area not covered by snow well into winter, which is typical of the dry postmonsoon period. (Google Earth.)

snowline has begun to rise to 5700 m. The rise has continued to 5750 by December 21. The lake at the terminus remains unfrozen.

10.2 South Lhonak Glacier

The South Lhonak Glacier drains east from Lhonak Peak into Sikkim from the Nepal – Sikkim border. The glacier begins at the border at 6500 m and ends in a lake at 5200 m. In 1933 the Mount Everest Expedition led by Eric Shipton explored the area. They traversed across the border and down the

Lhonak Glacier finding only a moraine-covered glacier and no lake. Raj *et al.* (2013) report on Lhonak Lake expanding by 1.9 km due to glacier retreat from 1962 to 2008.

In 1991 the glacier ended at the red arrow with a lake that is 1.0 km long, and the lowest 1 km of the glacier is heavily covered with debris. In 2015 the glacier has retreated 900 m since 1991. There is thinning evident at the purple arrow at a high altitude of 5800 m. A close-up examination of the terminus region of the South Lhonak Glacier indicates a well-incised supraglacial stream, indicating relative stagnation. The specific hazard posed by the glacier is the probability of a glacier lake outburst flood, which is a very high 42%, with a peak discharge estimated at 586 $m^3 s^{-1}$ (Raj *et al.*, 2013).

10.3 North Lhonak Glacier

This glacier is the exception in the northern part of this basin in that it lacks a proglacial lake at the terminus. The glacier is heavily covered with debris in the lower 1.5 km, and this combined with the lack of a lake has resulted in less retreat than nearby glaciers. The glacier had retreated 450 m from 1991 to 2015. The medial moraines have spread upglacier to 5700 m indicative of a rising equilibrium line, since medial moraines are not exposed above the equilibrium line on a glacier (purple arrow). The lower 1.5 km of the glacier remains stagnant with numerous supraglacial ponds and incised supraglacial streams (Fig. 10.4).

10.4 East Langpo Glacier

This glacier flows east from Langpo Peak. The ablation area of the glacier is heavily covered with debris. In 1991 the glacier terminus was in a small proglacial lake that was 500 m long and 100–200 m wide (Fig. 10.5). In 2015 the glacier had retreated 700 m, and the lake was now 1200 m long and widening to 400 m at the glacier terminus. The lower section of the glacier remains heavily covered with debris. The transient snowline during the October–December periods is typically at 5600 m.

10.5 Changsang Glacier

Changsang Glacier is a valley glacier just north of Kanchenjunga, the third highest peak in the world. A comparison of Landsat imageries from 1991 to 2015 identified the formation of a lake at the end of the glacier. In 1991 there was no evidence of a lake, and the glacier ended on an outwash plain at the red arrow. In 2000 there were a several small lakes beginning to develop. In 2011 the main lake was 1000 m long and has one debris-covered ridge that separates it from a second lake. In 2015 the lake was 1500 m and the glacier had retreated 1900 m in 24 years. The Changsang Glacier was reported to retreat 22 $m\,a^{-1}$ from 1976 to 2005 (Raina, 2009). From 1991 to 2015, the retreat rate has increased to 80 $m\,a^{-1}$.

A close-up view of the terminus area in 2006 indicates the main lake and several smaller lakes that joined the main lake in 2012. The purple arrow indicates the outlet river from beneath the stagnant debris-covered ice. The orange arrows indicate the extent of the developing lake in 2012 (Fig. 10.6). The ice in the lake is dead ice, which will melt faster than the rest of the debris-covered glacier (Watanabe, Kameyama, and Sato, 1995).

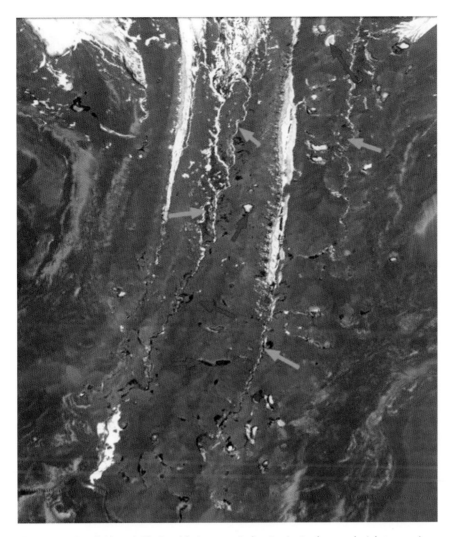

Figure 10.4 North Lhonak Glacier ablation zone indicating incised supraglacial streams (green arrows) and supraglacial ponds (pink arrows).

10.6 Zemu Glacier

Zemu Glacier is a 26 km long glacier draining the east side of Kanchenjunga, the world's third highest mountain (Fig. 10.7). The importance of the glacier is that it is a key water source for the Teesta River. The glacier acts as a natural reservoir releasing water due to melting. The Teesta River is the focus of a hydropower development project being undertaken by the Government of Sikkim. To date 510 MW of the proposed 3500 MW potential is operating. This is a run-of-the-river project, with the water extracted from the river without a dam, run along the valley wall, and dropped back to the river through a series of turbines. Run of river is much less expensive than a dam in this remote, earthquake-prone mountainous valley.

Figure 10.5 Changes in Jongsang (J), East Langpo (EL), and Changsang (C) Glaciers from 1991 to 2015. Red arrow indicates the 1991 terminus, yellow arrow the 2015 terminus, and purple arrow areas of thinning. Green arrow indicates a new lake forming.

Figure 10.6 Google Earth imagery from 2006, indicating collapse of stagnant ice in proglacial lake (orange arrows). Purple arrow indicates outlet stream. (Google Earth.)

Near the head of the glacier, the walls of Kanchenjunga deliver the debris and large amounts of snow in the form of avalanches to the glacier basin at 5900–5200 m. The lower 18 km of the glacier is in the ablation zone where melt dominates. The glacier has a heavy debris cover on most of its length, insulating it from ablation and leading to no detectable retreat of the main terminus from 2000 to 2013 (Basnett, Kulkarni, and Bolch, 2013). A view of the lower glacier indicates this heavy debris

Figure 10.7 Changes in Zemu (Z) and Hidden (H) Glaciers from 1991 to 2015. Red arrow indicates the 1991 terminus, and purple arrow indicates areas of thinning and retreat of tributary glaciers.

cover, with some scattered small glacial lakes on its surface. Several of the tributaries no longer join the Zemu, depriving it of a portion of its former accumulation sources. Purple arrows indicate three glaciers in the watershed that have all been slowly retreating, indicating that there is less contribution to the lower glacier from higher elevations.

Thinning has been significant. The lateral moraine ridges on either side of the main glacier average 50 m above the main glacier surface. These were built during the Little Ice Age advance. Lateral moraines do not reach above the glacier surface that built them. Thus, the lower glacier has thinned by approximately 50 m in approximately the last century.

A view of a portion of the upper glacier indicates one issue for the glacier. Several of the tributaries no longer join the Zemu, depriving it of a portion of its former accumulation sources. A comparison of 2000 and 2013 Landsat images indicates the lack of change in the location of the main terminus (red arrows) and recession of surrounding glaciers in the Zemu Basin (yellow arrows).

This area from 5200 to 4200 m terminus would quickly melt away without the natural debris cover. The glacier receives considerable snow input from up to 8000 m via avalanches, which are deposited in this region between 5200 and 5900 m. This glacier will continue to be a large water source for the Teesta River for the foreseeable future. The glacier has not been retreating as fast or developing a proglacial lake as has happened at most other glaciers in the area. This should be anticipated in the near future.

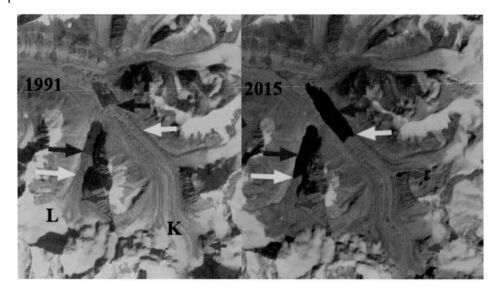

Figure 10.8 Changes in Kaer (K) and Longbashaba (L) Glaciers from 1991 to 2015. Red arrow indicates the 1991 terminus, yellow arrow the 2015 terminus, and purple arrow areas of thinning.

10.7 Kaer Glacier

A comparison of imageries from 1991 to 2015 indicates an expansion of lakes at the terminus of Kaer Glacier and Longbashaba Glacier (Fig. 10.8). Yao *et al.* (2012) identified the lake at the terminus and the glacier as Longbashaba; however, GLIMS and the Chinese Glacier inventory of 2010 indicated this as Kaer Glacier and the lake is unnamed even in the detailed Pumqu Basin inventory (Che *et al.*, 2014). Here we use the inventory labeling. Kaer Glacier is at the headwaters of the Yairuzangbo river in the Pumqu Basin. Kaer Glacier terminates in a substantial moraine-dammed lake in Tibet (Yao *et al.*, 2012). The concern is to the 23 villages downstream of the lake and the Rongkong Hydropower station. The lake was quite small in 1991 with a length of 900 m. The lake expanded throughout the period to 2015 to a length of 2200 m, with glacier retreat of 1300 m during the period. There is noticeable expansion of bare rock areas upglacier, indicating reduced accumulation at higher elevations that will lead to less ice being communicated to the terminus. The lake is showing no signs of narrowing, suggesting the upvalley limit is not being approached. As the lake depth has increased, the volume has quadrupled to 1.3 km^3 (Yao *et al.*, 2012).

10.8 Longbashaba Glacier

Longbashaba Glacier ends in a proglacial lake that nearly joins the lake at the end of Kaer Glacier. The lake is unnamed in the Pumqu Basin inventory of glacier lakes, but Yao *et al.* (2012) indicated this to be the Pida Lake. In 1991 the lake was 1200 m long. The lake has a consistent width of 400 m and has expanded to a length of 2.1 km in 2015. The 900 m retreat will continue as indicated by reduced ice flow upglacier. The purple arrow 1 km above the terminus indicates considerable thinning since 1991. The upvalley limit of this lake has not been approached.

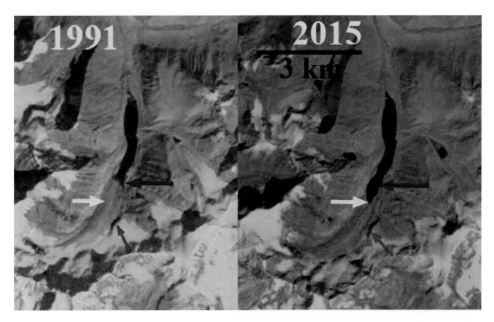

Figure 10.9 Changes in Zhizhai Glacier from 1991 to 2015. Red arrow indicates the 1991 terminus, yellow arrow the 2015 terminus, and purple arrow areas of thinning.

10.9 Zhizhai Glacier

Located between Longbashaba Glacier and Jimi Glacier, this glacier extends 4.75 km north from the Nepal–China border in the southeast region of the Pumqu River Basin. The moraine-dammed glacier lake, Zhuxico Lake, poses some risk to the 23 villages downstream of the lake and the Rongkong Hydropower station. This risk has been examined for another lake in the Longbashaba Basin by Yao *et al.* (2012). Che *et al.* (2014) report on an inventory of glaciers and glacier lakes in the Pumqu Basin and note that lake expansion is higher from 2001 to 2013 than for the 1970–2000 period, posing greater threats for a glacier lake outburst flood. They report that there are 254 glaciers lakes in the basin, currently 55 of which have formed since 1970. In 1991 the glacier terminated in a proglacial lake that was 2.25 km long (Fig. 10.9). A tributary flowed around a ridge and rejoined the main glacier (purple arrow). In 2000 the glacier had retreated 200 m, the tributary at the purple arrow remains connected to the main glacier. In 2015 the glacier had retreated 600 m and the lake was 2.9 km long and remained 300 m wide. The tributary at the purple arrow no longer rejoins the main glacier. Just east of Zhizhai Glacier, an unnamed glacier also features an expanding lake due to glacier retreat.

10.10 Jimi Glacier

Jimi Glacier terminates in Gyemico Lake, a moraine-dammed lake at 5100 m. In 1991 the glacier ended at the red arrow in a section of the lake that was 400 m wide (Fig. 10.10). In 2015 the glacier had retreated 500 m and terminated in a narrower section of the lake, 250 m; it appears that the glacier is nearing the head of the lake.

Figure 10.10 Changes in Jimi (J) and Yindapu (Y) Glaciers from 1991 to 2015. Red arrow indicates the 1991 terminus, yellow arrow 2015 terminus, and purple arrow areas of thinning.

10.11 Yindapu Glacier

Yindapu Glacier has a debris-covered terminus and ends near the shore of an unnamed glacier-dammed lake. In 1991 the glacier had retreated from this lake. From 1991 to 2015, there is not a measurable retreat of the glacier. The debris cover appears to be from a landslide, as the rest of the glacier is quite clean; this has helped insulate the terminus reducing retreat (Fig. 10.11).

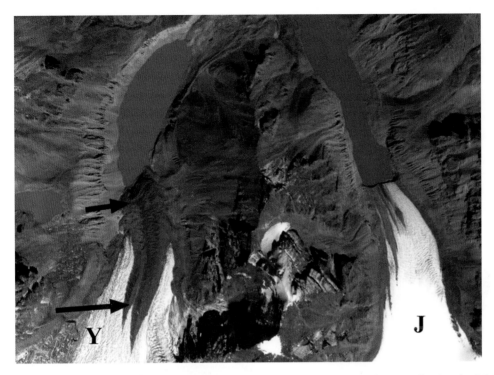

Figure 10.11 Google Earth image of the terminus of Jimi (J) and Yindapu (Y) Glaciers, indicating the debris cover increase on Yindapu Glacier. (Google Earth.)

10.12 Gelhaipuco Glacier

Gelhaipuco Glacier is a glacier-dammed lake at the headwaters of the Natangqu River. In 1964 the lake had an outburst flood that resulted in severe damage, and heavy economic losses occurred in the Chinese territory and downstream in the Arun Valley in Nepal. The glacier that ends in it is unnamed but is referred to here by the name of the lake. In 1991 the glacier terminated at the red arrow in the lake, which was 750 m long (Fig. 10.12). In 2015 glacier retreat had expanded the lake to 1500 m. The glacier retreat of 800 m is occurring in a lake that is maintaining consistent width. The lake has an estimated volume of 25 million m³ and is a risk for a glacier outburst flood (Che *et al.*, 2014). The lake is dammed by a moraine dam that has failed (Fig. 10.13).

10.13 Qangzonkco Glacier

Qangzonkco Glacier is a glacier-dammed lake near the headwaters of the Natangqu River. Here we refer to the glacier that ends in the lake and is unnamed with the lake name. In 1991 the glacier ended at the red arrow and the lake was 1700 m long. In 2015 the glacier had retreated 900 m and the lake was 2600 m long. The lake has an estimated volume of and is a risk for a glacier outburst flood. The lake is narrowing with glacier retreat.

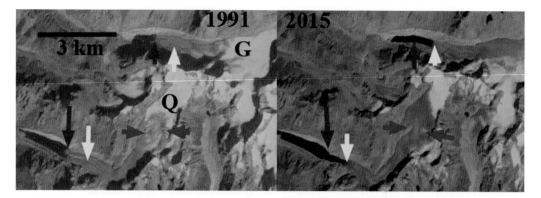

Figure 10.12 Changes in Gelhaipuco (G) and Qangzonkco (Q) Glaciers from 1991 to 2015. Red arrow indicates the 1991 terminus, yellow arrow the 2015 terminus, and purple arrow areas of thinning.

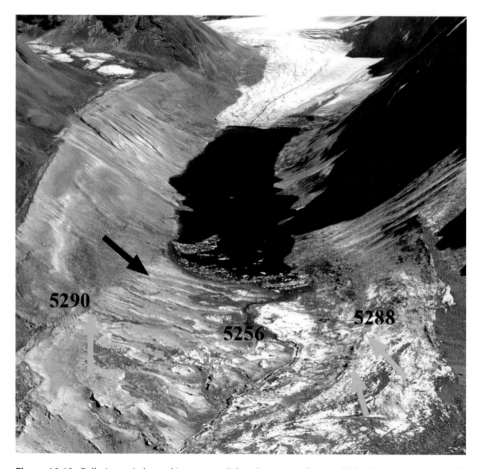

Figure 10.13 Gelhaipuco Lake and its unconsolidated moraine-dammed lake. Note the elevation listed near the former shoreline and the current outlet stream.

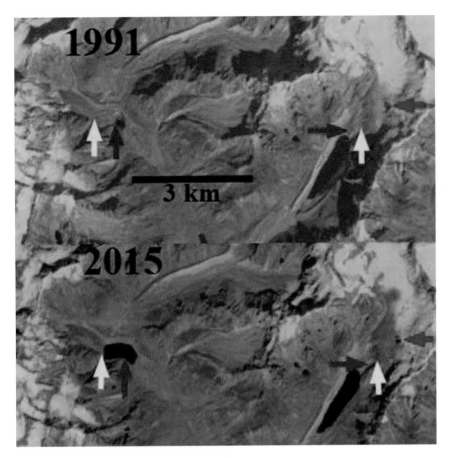

Figure 10.14 Changes in Nobuk (N) and Nangama Pokhari (Np) Glaciers from 1991 to 2015. Red arrow indicates the 1991 terminus, yellow arrow 2015 terminus, and purple arrow areas of thinning.

10.14 Nobuk Glacier

At the headwaters of the Tamor Basin in eastern Nepal is an unnamed glacier that terminates in an expanding glacial lake (Fig. 10.14). The glacier is referred to as "Nobuk" Glacier here in reference to the nearby named peak. The glacier is upstream of Chheche Pokhari, a lake formed by a glacier outburst flood in 1980. Two arms of the glacier are avalanche-fed from the steep border peaks with Tibet. The ICIMOD has recently finished a detailed inventory of glacier change in Nepal since 1980. In the Tamor Basin they indicate that the glacier area from 2000 to 2010 has declined from 422 to 386 km^2 (Baidya, 2014).

Here the glacier is examined from 1991 to 2015 using Landsat images. In 1991 the lake had several developing areas amidst the decaying glacier ice, but the glacier still reached to the far eastern shore of the lake, with a maximum lake width of 500 m. The glacier terminus is fed by two arms, with the southern arm having a steep icefall near the terminus, and is narrower. In 2000 Nobuk Glacier terminated at a southeast turn on the south side of the glacial lake it terminates in (red arrow). In 2009 the two arms of the glacier were separated, and the southern arm no longer reached the lake. The glacier front has retreated back to the base of the buttress at Point A. In 2015 the lake has more

Figure 10.15 Google Earth image 2010 indicates the former terminus (red arrows) and current terminus (green arrows) of Nobuk Glacier and the lateral moraine adjacent to the lake. (Google Earth.)

than doubled in length and area since 2000. The southern arm terminates 150 m from the lake, and the debris-covered northern arm, though still ending in the lake, is a very thin low slope terminus that appears to be close to retreating from the lake basin that the glacier has carved. This is evident in the 2010 Google Earth image (Fig. 10.15). The glacier had retreated 700 m from 1991 to 2015. The lake is now 1 km long and has an area of 0.4 km^2. The proglacial lake is dammed by a moraine of unknown stability.

10.15 Nangama Pokhari

Nangama Pokhari is near the headwaters of the Tamor Basin. In 1991 the glacier terminus was near the northern shore of a glacial lake impounded by a glacier moraine (Fig. 10.14). In 2015 the glacier had retreated 500 m from the lakeshore. The glacier has significant upglacier thinning that will lead to continued retreat.

10.16 Kanchenjunga Glacier

Kanchenjunga Glacier is the main glacier draining west from Kanchenjunga Peak, also listed on maps as Kumbukarni (Fig. 10.16). The glacier is similar to Zemu Glacier flowing east from the same mountain into Sikkim, in the heavy debris cover that dominates the glacier in the ablation zone extending from the terminus for 15 km and an altitude of 5600 m (Racoviteanu and Williams, 2012). Identifying the retreat is difficult due to the debris cover. Racoviteanu *et al.* (2015) examined glaciers in this region

Figure 10.16 Kanchenjunga Glacier (K) from 1991 to 2015. Green arrows indicate locations of enhanced supraglacial lakes since 1991. Purple arrow indicates areas of thinning at higher elevations in the region.

using 1962 and 2000 images. They found area losses of 14% for debris-covered glacier and 34% for clean glaciers. The debris-covered glaciers' terminus response is even more muted, indicating why terminus change is an easy measure of glacier change, but not always the best. For Kanchenjunga Glacier, Racoviteanu *et al.* (2015) indicated that the glacier area declined by just 4–8% from 1962 to 2000.

What is apparent in the Landsat images at the green arrows is the increase from 1991 to 2015 of supraglacial lakes. Also features of thinning are evident in the midreaches of the glacier (purple arrows), where tributaries have narrowed and detached from the main glacier. A close-up of the main glacier junction 12 km above the terminus indicates the number of large supraglacial lakes (Fig. 10.17). These cannot form in a region where melting does not dominate over glacier motion. The Google Earth image from 2014 of the terminus area indicates a patchwork of moraine-cored ice dotted with supraglacial lakes and dissected by the glacial outlet river in the lower 3 km of the glacier (Fig. 10.18). This is clearly not an active portion of the glacier that is thin stagnant and does not fill even the valley floor. An overlay of images indicates the lack of motion. The heavy debris cover has slowed retreat

Figure 10.17 Supraglacial lakes 12 km above the terminus at the main junction region at 5200 m.

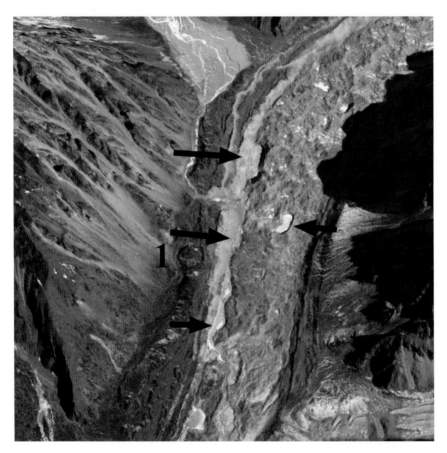

Figure 10.18 Marginal thinning and a supraglacial lake 2–3 km above the terminus at 4800 m.

and thinning; however, the lower glacier is poised for an increased rate of retreat with merging of supraglacial lakes, which will lead to further area losses.

References

Ageta, Y. and Higuchi, K. (1984) Estimation of mass balance components of a summer-accumulation type glacier in the Nepal Himalaya. *Geografiska Annaler: Series A, Physical Geography*, **66** (3), 249–255.

Baidya, S. (2014) *Glacier status in Nepal and decadal change from 1980 to 2010 based on landsat data*, ICIMOD, Kathmandu http://lib.icimod.org/record/29591/files/GSN-RR14-2.pdf.

Bajracharya, S.R., Maharjan, S.B., and Shresta, F. (2014) The status and decadal change of glaciers in Bhutan from 1980's to 2010 based on the satellite data. *Annals of Glaciology*, **55**, 159–166. doi: 10.3189/2014AoG66A125

Bajracharya, S.R. and Mool, P. (2009) Glaciers, glacial lakes and glacial lake outburst floods in the Mount Everest region, Nepal. *Annals of Glaciology*, **50** (53), 81–86. doi: 10.3189/172756410790595895

Bajracharya, S.R., Mool, P.K., and Shrestha, B.R. (2007) *Impact of Climate Change on Himalayan Glaciers and Glacial Lakes*, ICIMOD, Kathmandu, Nepal, p. 119.

Bajracharya, S.R. and Shrestha, B. (2011) *The Status of Glaciers in the Hindu-Kush Himalayan Region*, International Centre for Integrated Mountain Development (ICIMOD), Kathmandu, Nepal, p. 127.

Basnett, S., Kulkarni, A., and Bolch, T. (2013) The influence of debris cover and glacial lakes on the recession of glaciers in Sikkim Himalaya, India. *Journal of Glaciology*, **59**, 1035–1046.

Bhambri, R., Bolch, T., Chaujar, R.K., and Kulshreshtha, S.C. (2011) Glacier changes in the Garhwal Himalaya, India, from 1968 to 2006 based on remote sensing. *Journal of Glaciology*, **57** (203), 543–556.

Bolch, T., Kang, Y.S., Buchroithner, M.F. *et al.* (2010) Glacier inventory for the western Nyainqentanglha Range and the Nam Co Basin, Tibet, and glacier changes 1976–2009. *The Cryosphere*, **4**, 419–433.

Bolch, T., Kulkarni, A., Kääb, A. *et al.* (2012) The state and fate of Himalayan glaciers. *Science*, **336** (6079), 310–314. doi: 10.1126/science.1215828

Bolch, T., Pieczonka, T., and Benn, D.I. (2011) Multidecadal mass loss of glaciers in the Everest area (Nepal Himalaya) derived from stereo imagery. *The Cryosphere*, **5**, 349–358.

Che, T., Xiao, L., and Liou, Y. (2014) Changes in Glaciers and Glacial Lakes and the Identification of Dangerous Glacial Lakes in the Pumqu River Basin, Xizang (Tibet). *Advances in Meteorology*, **2014**, Article ID 903709. doi: 10.1155/2014/903709

ICIMOD (2011) *Glacial Lakes and Glacial Lake Outburst Floods in Nepal*, ICIMOD, Kathmandu, Nepal.

ICIMOD (2014) *Glacier Dynamics in Nepal*, http://apps.geoportal.icimod.org/nepalglacier/ (accessed 10 February 2015).

Immerzeel, W.W., Beek, L.P., and Bierkens, M.F. (2010) Climate change will affect the asian water towers. *Science*, **328**, 1382. doi: 10.1126/science. 1183188

Jacob, T., Wahr, J., Pfeffer, T., and Swenson, S. (2012) Recent contributions of glaciers and ice caps to sea level rise. *Nature*, **482**, 514–518. doi: 10.1038/nature10847

Jin, R., Li, X., Che, T. *et al.* (2005) Glacier area changes in the Pumqu river basin, Tibetan Plateau, between the 1970 s and 2001. *Journal of Glaciology*, **51** (175), 607–610.

Kehrwald, N.M., Thompson, G., Tandong, L. *et al.* (2008) Mass loss on Himalayan glacier endangers water resources. *Geophysical Research Letters*, **35**, L22503. doi: 10.1029/2008GL035556

Kulkarni, A.V., Bahuguna, I.M., Rathore, B.P. *et al.* (2007) Glacial retreat in Himalaya using Indian remote sensing satellite data. *Current Science*, **92** (1), 69–74.

Kulkarni, A. and Karyakarte, Y. (2014) Observed changes in Himalayan glaciers. *Current Science*, **106** (2), 237–244.

Li, Z.G., Yao, T.D., Tian, L.D. *et al.* (2011) Glacier in the upstream Manla Reservoir in the Nianchu River Basin, Tibet: Shrinkage and impacts. *Sciences in Cold and Arid Regions*, **3** (2), 0110–0118.

Matsuo, K. and Heki, K. (2010) Time-variable ice loss in Asian high mountains from satellite gravimetry. *Earth and Planetary Science Letters*, **290**, 30–36. doi: 10.1016/j.epsl.2009.11.053

Neckel, N., Kropacek, J., Bolch, T., and Hochschild, V. (2014) Glacier mass changes on the Tibetan Plateau 2003–2009 derived from ICESat laser altimetry measurements. *Environmental Research Letters*, **9**, 014009. doi: 10.1088/1748-9326/9/1/014009

Pelto, M. (2010) Forecasting temperate Alpine Glacier survival from accumulation zone observations. *The Cryosphere*, **3**, 323–350.

Pelto, M. (2011) Hydropower: hydroelectric power generation from alpine glacier melt, in *Encyclopedia of Snow, Ice and Glaciers* (eds V.P. Singh, P. Singh, and U.K. Haritashya), Springer, Dordrecht.

Racoviteanu, A.E., Arnaud, Y., Williams, M.W., and Manley, W.F. (2015) Spatial patterns in glacier characteristics and area changes from 1962 to 2006 in the Kanchenjunga–Sikkim area, eastern Himalaya. *The Cryosphere*, **9**, 505–523. doi: 10.5194/tc-9-505-2015

Racoviteanu, A. and Williams, M.W. (2012) Decision tree and texture analysis for mapping debris-covered glaciers in the Kangchenjunga area, Eastern Himalaya. *Remote Sensing*, **4** (10), 3078–3109.

Raina, V.K. (2009) *Himalayan Glaciers: A State-of-Art Review of Glacial Studies, Glacial Retreat and Climate Change (Discussion Paper)*, Ministry of Environment and Forests, Government of India, New Delhi, 56 pp, http://www.moef.nic.in/downloads/public-information/MoEF%20Discussion%20Paper%20_him.pdf Discussion Pape_him.pdf.

Raj, G.K.B., Remya, S., and Kumar, K.V. (2013) Remote sensing-based hazard assessment of glacial lakes in Sikkim Himalaya. *Current Science*, **104**, 359–363.

Watanabe, T., Kameyama, S., and Sato, T. (1995) Imja Glacier dead-ice melt rates and changes in a supraglacial lake, 1989–1994, Khumbu Himal, Nepal: danger of lake drainage. *Mountain Research and Development*, **15** (4), 293–300.

WWF (2005) *An Overview of Glaciers, Glacier Retreat, and Subsequent Impacts in Nepal, India and China (WWF Nepal Program)*, World Wildlife Fund, Washington, DC, p. 79.

Xu, J., Shrestha, A., and Eriksson, M. (2009) Climate change and its impacts on glaciers and water resource management in the Himalayan Region. *Assessment of Snow, Glacier and Water Resources in Asia*. International Hydrological Programme of UNESCO and Hydrology and Water Resources Programme of WMO.

Yao, T., Thompson, L., Yang, W. *et al.* (2012) Different glacier status with atmospheric circulations in Tibetan Plateau and surroundings. *Nature Climate Change*, **2** (8), 663–667. doi: 10.1038/nclimate1580

Zhou, C., Yang, W., Wu, L. *et al.* (2009) Glacier changes from a new inventory, Nianchu river basin, Tibetan Plateau. *Annals of Glaciology*, **50** (53), 87–92.

11

New Zealand

Overview

The Southern Alps of New Zealand are host to over 3000 glaciers that owe their existence to high amounts of precipitation ranging from 3 to 10 m (Chinn, 1999). The NIWA glacier monitoring program has noted that the volume of ice in New Zealand's Southern Alps has decreased by 36% with the loss of 19.0 km^3 of glacier ice from 53.3 km^3 in 1978 to 34.3 km^3 in 2014 (New Zealand Government, 2015). Volume loss in New Zealand glaciers is dominated by 12 large glaciers (Salinger and Willsman, 2007). More than 90% of this loss is from 12 of the largest glaciers in response to rising temperatures over the twentieth century (Chinn, 1999). From 1977 to 2015, NIWA has conducted an annual snow line survey; in 6 of the last 9 years, the snow line has been significantly above average and 3 years approximately at the average (Willsman, Chinn, and Lorrey, 2014). This has driven the widespread glacier retreat underway. In each case the retreat of the largest glaciers has been enhanced by the formation and expansion of lakes in this newly developing lake district. Dykes *et al.* (2011) identify the role of glacier lakes in accelerating the retreat of Tasman Glacier. The retreat of these glaciers has until recently been slowed by debris cover and the long low slope ablation zone segments (Chinn, 1996; Chinn, 1999). Glaciers that lack debris cover and have a steeper slope have a more rapid response time, such as Fox and Franz Josef Glaciers (Purdie *et al.*, 2014). In the 1972 map of the region, there is no lake at the terminus of the Tasman, Mueller, or Hooker Glaciers where each has a substantial lake in 2015 (Fig. 11.1). Each lake continues to expand as glacier retreat continues.

Lake Tekapo and Lake Pukaki are both utilized for hydropower. Hooker, Mueller, Murchison, and Tasman Glaciers drain into Lake Pukaki, where water level has been raised 9 m for hydropower purposes. Water from Lake Pukaki is sent through a canal into the Lake Ohau watershed and then through six hydropower plants of the Waitaki hydro scheme: Aviemore, Benmore, Waitaki, and Ohau A, B, and C with a combined output of 1340 MW. Meridian owns and operates all six hydro stations located from Lake Pukaki to Waitaki. The reduction of glacier area in the region will reduce summer runoff into Lake Pukaki and this hydropower system.

11.1 Mueller Glacier

Mueller Glacier drains the eastern side of Mount Sefton, Mount Thompson, and Mount Isabel. The lower section of the glacier is debris covered in the valley reach from the terminus at 1000–1250 m. The lake at the end of the Mueller Glacier was just forming with a length of less than 400 m in 1990 (Fig. 11.2). In 2004 the Mueller Glacier Lake had expanded to a length of 700 m. In 2015 the lake had reached 1800 m in length. Mueller Lake had a surface area of 0.87 km^2 and a maximum depth

Recent Climate Change Impacts on Mountain Glaciers, First Edition. Mauri Pelto.
© 2017 John Wiley & Sons, Ltd. Published 2017 by John Wiley & Sons, Ltd.

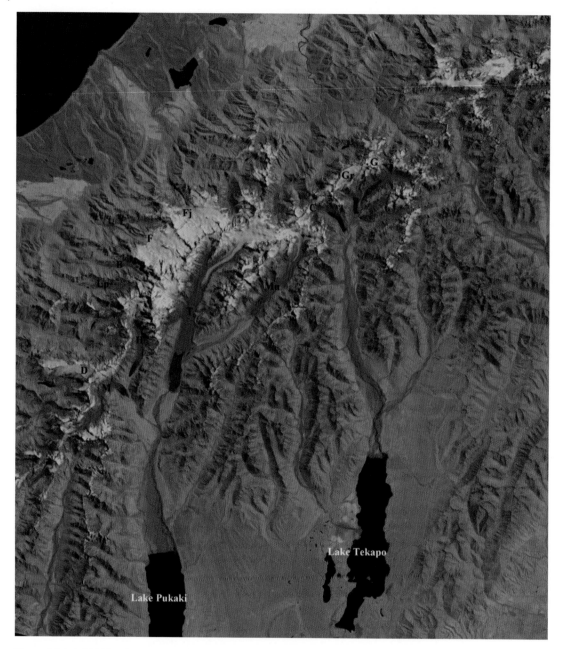

Figure 11.1 A 2015 Landsat image indicating New Zealand glaciers examined. D = Douglas, Lp = La Perouse, and B = Balfour, the development since 1972 of lakes at the end of Hooker, Mueller, and Tasman Glaciers.

Figure 11.2 Glacier change revealed in Landsat images of Mueller (M) and Hooker (H) Glaciers from 1990 and 2015. The red arrows indicate 1990 terminus location, the yellow arrows 2015 terminus location, and the purple arrows upglacier thinning.

of 83 m (Robertson *et al.*, 2012). The 1500 m retreat from 1990–2015 will continue in the future as the lower 2 km section of the glacier is stagnant. The stagnant nature is evident from the formation of supraglacial lakes and considerable surface roughness (green arrow) that does not occur when a glacier is active and moving. The glacier has been fed by three different tributary glaciers flowing off of Mount Sefton. Two of them, Tuckett and Huddleston, are no longer delivering significant ice to the Mueller; only modest avalanching now spills onto the Mueller Glacier. Only the Frind Glacier is currently contributing to the Mueller Glacier. The result is that the end of truly active ice is several kilometers upglacier of the present terminus; this will develop into the terminus of the Mueller Glacier. The lake has not been surveyed, but seems to lack the depth at the current terminus of Tasman Lake where calving can be more important. Robertson *et al.* (2012) have noted that subaqueous calving has been an important process of ice loss for the glacier.

11.2   Hooker Glacier

Hooker Glacier parallels the Tasman Glacier one valley to the west draining south from Mount Hicks and Mount Cook. Hooker Glacier is a low-gradient glacier, which helps reduce its overall velocity and a debris-covered ablation zone reducing ablation, both factors increasing response time to climate change (Quincey and Glasser, 2009). Hooker Lake, which the glacier ends in, began to from around 1982 (Kirkbride, 1993). In 1990 the lake was 1100 m long (Fig. 11.2). From 1990 to 2015 the lake expanded to 2300 m, with the retreat enhanced by calving. The 1200 m retreat was faster during the earlier part of this period (Robertson et al., 2013). The lower 3.4 km of the glacier has limited motion. Robertson et al. (2013) suggest that the retreat will end after a further retreat of 700–1000 m as calving will decline as the lake depth declines. The peak lake depth is over 130 m, with the terminus moving into shallow water after 2006 leading to declining retreat rates (Robertson et al., 2012). Based on the nearly stagnant nature of the lower glacier and the diminished ice flow from above indicated by debris cover expansion at the purple arrow, it seems likely that the retreat will continue well beyond the end of the lake but at a diminished rate.

11.3   Tasman Glacier

Tasman Glacier flows south from the highest portion of the Southern Alps on the east side of Mount Cook and Mount Tasman. The glacier ends in the expanding Tasman Lake (Fig. 11.3). The glacier is the longest, largest glacier in New Zealand with a low-gradient and debris-covered ablation zone. The combination of these three leads to the glacier having a longer response time to climate change (Quincey and Glasser, 2009). Dykes et al. (2011) note a retreat of 180 m a^{-1} since the 1990s. In the 1972 map of the region, there is no lake at the terminus of the glacier. In 2000 the Tasman Lake was 2 km long on the west side and 4.5 km along on the east side. In 2006 the lake has expanded 2.5 km on the west side and 5.5 km long in a narrow tongue on the east side. Tasman Lake is now 7 km long across a width of about 2 km. Dykes et al. (2011) note a maximum depth of 240 m and an expansion of 0.34 km^2 a^{-1} in area to a total lake area of 6 km^2 in 2010. The proglacial lake at the terminus continues to expand as the glacier retreats upvalley. The lake is deep with most of the lake exceeding 100 m in depth, and the valley has little gradient, thus the retreat will continue; the increasing lake depth has been key to increased calving and retreat (Fig. 11.4) (Dykes and Brook, 2010). The lake can expand in this low-elevation valley another 9 km, and at the current rate, this will occur over two decades (Dykes et al., 2011).

The glacier has experienced two larger calving events in recent years: the first triggered by the Christchurch earthquake in February 2011 and the second on January 30, 2012. Such events can occur because the terminus has thinned to the point that the glacier terminus is more buoyant and crevasses and rifts extend through the thinner ice. Subaerial calving results largely from development of a waterline notch, which is facilitated by constant lake water levels (Rohl, 2006). Subaqueous calving also occurs on Tasman Glacier from ice ramps extending beneath the lake surface (Rohl, 2006). The retreat is enhanced by calving into the lake, but most of the volume loss, which average −0.87 m w.e. a^{-1} from 1986 to 2008, is from downwasting of the glacier.

11.4   Murchison Glacier

Murchison Glacier drains south in the next valley east of Tasman Glacier and terminates in a lake that is rapidly developing as the glacier retreats. This lower section is debris covered, stagnant, relatively flat, and will not survive long. There was not a lake in the 1972 map of the region. In 1990 the newly formed lake was limited to the southeast margin of the terminus (Fig. 11.5). From 1990 to 2015 the terminus has retreated 2700 m. A rapid retreat will continue, as the 2010, 2013, and 2015 images indicate that other proglacial lakes have now developed 3.5 km above the actual terminus. These lakes

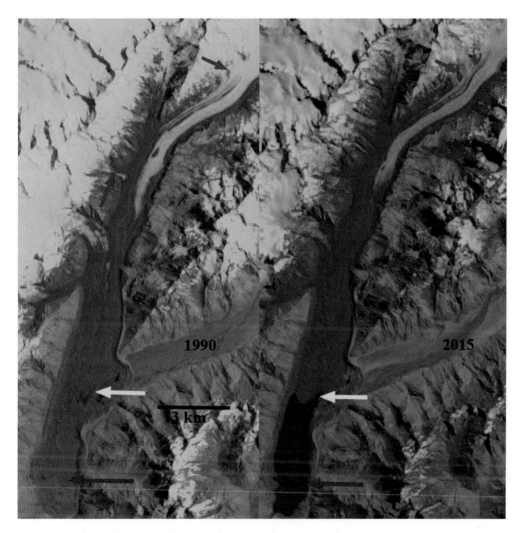

Figure 11.3 Glacier change revealed in Landsat images from 1990 and 2015 at Tasman Glacier. Red arrows indicate 1990 terminus location, the yellow arrows 2015 terminus location, and the purple arrows upglacier thinning.

are at a higher elevation and may not endure but do help increase ablation, and in Fig. 11.5 show a glacier that is too narrow to provide flow to the lower 3.5 km. The demise of the lower section of this glacier will parallel to that of Tasman Glacier. The increased retreat has been forecast by the NIWA and Dykes and Brook (2010).

The February 2011 earthquake near Christchurch led to a major calving event of a portion of the stagnant terminus reach of the Tasman Glacier. There is no evident calving event from Murchison Glacier. The lake on the western margin of the valley, separated from the main lake, has since April 2010 expanded notably.

The glacier still has a significant accumulation area above 1650 m to survive at a smaller size. The lower debris-covered tongue is 6 km long and extends from the terminus at 1050–1200 m, a very low gradient to supply healthy flow from the accumulation area. The ongoing retreat is triggered by warming and a rise in the snow line in the New Zealand Alps observed by the Willsman, Chinn, and

Figure 11.4 Tasman Glacier terminus with supraglacial lakes in a 2013 Google Earth image indicated by blue arrows and icebergs from calving with black arrows. (Google Earth.)

Lorrey (2014). This has led to increased exposure of bedrock high on the glacier and reduction of tributary inflow noted at purple arrows.

11.5 Douglas Neve

The Douglas Neve flows down the steep side of Mount Scott and Seddon Peak. The bedrock slope at the base of the glacier is particularly smooth, which combined with the steep slope, 40% grade or 22°

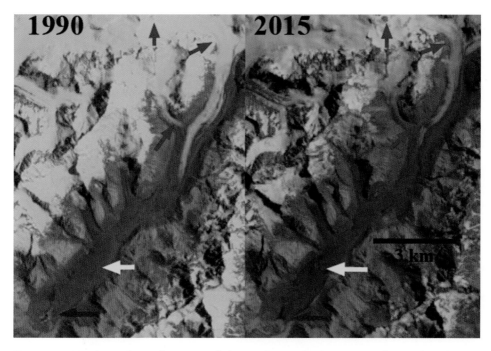

Figure 11.5 Murchison Glacier change revealed in Landsat images from 1990 and 2015. The red arrows indicate 1990 terminus location, the yellow arrows 2015 terminus location, and the purple arrows upglacier thinning.

slope, enhances basal sliding. On small alpine glacier, the resulting high velocity leads to extensive crevassing. This crevassing can literally penetrate to the base of the glacier near the thin terminus. This leads to portions of the glacier simply separating from the rest of the glacier and avalanching down the slope or melting in place.

Here we utilize Landsat images from 1990 to 2015 and Google Earth imagery from 2004 to 2009 to examine the retreat of this glacier (Fig. 11.6). In 1990 the terminus of the glacier terminated on the west at the red dot and the east at the red arrow. The glacier was still advancing reaching a maximum between 2000 and 2004 at a prominent bedrock fracture at 1640 m above sea level. The green line in the Google Earth imagery is the 2004 terminus and the burgundy line the 2009 terminus (Fig. 11.7). In 2009 the terminus has retreated 400 m and consists of two unsustainable narrow tongues, both less than 100 m side. In 2012 the two narrow tongues have been lost, resulting in a 700 m retreat from 2000 to 2012 with the terminus now at 1800 m. As the retreat of an alpine glacier progresses, crevassing typically is reduced as glacier speed declines. Here we see an increase in crevassing from 2004 (above) to 2009 (below) in the terminus area, suggesting that the retreat will continue via pieces of the glacier separating from the glacier and avalanching. This process is a much different setting but similar in practice to ice shelf loss through rifting that reaches the critical point where the rifts lead to icebergs breaking off. At this point the terminus remains unsustainable.

11.6   La Perouse Glacier

La Perouse Glacier drains west from Mount Hicks terminating at the head of the Cook River. Gjermundsen *et al.* (2011) noted that the glacier retreated 1900 m from 1978 to 2002. From

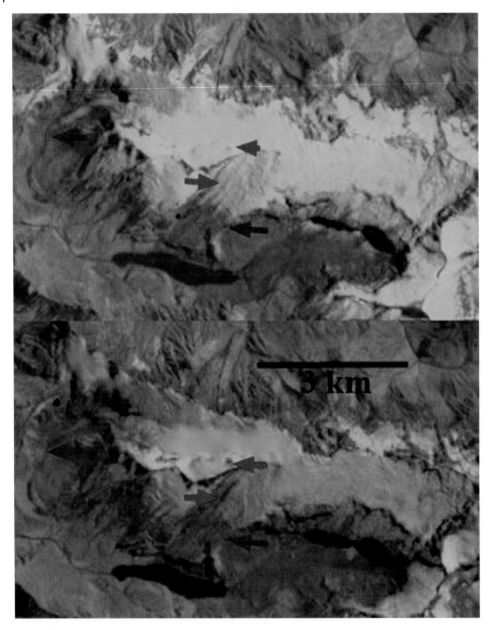

Figure 11.6 Glacier change revealed in Landsat images of Douglas Neve from 1990 and 2015. The red arrows indicate 1990 terminus location and the purple arrows upglacier thinning.

1990 to 2015, the retreat of the heavily debris-covered terminus has been limited (Fig. 11.8). The expansion of debris-covered upglacier and the development of proglacial lakes along the southern margin of the glacier indicate continued downwasting that will lead to a further period of rapid retreat.

Figure 11.7 Google Earth image from 2004 and 2009, with the green line from 2004 and the red line from 2009. (Google Earth.)

Figure 11.8 Glacier change revealed in Landsat images of La Perouse Glacier from 1990 and 2015. The red arrows indicate 1990 terminus location and the purple arrows upglacier thinning. The orange arrows indicate a region of decaying ice beyond the terminus that is being revegetated.

11.7 Balfour Glacier

Balfour Glacier drains west from Mount Tasman in the Southern Alps of New Zealand. The ablation area is a low slope, 8 km long debris-covered tongue extending from the terminus near 800–1600 m. The glacier is fed by avalanching off of Mount Tasman to the west, the southern flank of the Fox Range to the north, and the northern flank of the Balfour Range to the south. Gjermundsen *et al.* (2011) examined the change in glacier area in the central Southern Alps and found a 17% reduction in area mainly from reductions of large valley glaciers such as Balfour Glacier. In 1990 the glacier ended at 700 m with a snowline at 1600 m (Fig. 11.9). The lower 18 km of the Balfour Glacier is debris covered. Only the upper 8 km has a snow cover. In 2015 the terminus has retreated 1250 m, and the snow line is at 1800 m, with the lower 20 km debris covered. The terminus reach has continued to appear stagnant from 1990 to 2015. Balfour Glacier has not developed a significant proglacial lake at its terminus, which has limited the retreat compared with Tasman Glacier or Mueller Glacier. A Google Earth image indicates the retreat of stagnant debris. The primary glacier meltwater outlet issues from the glacier at the yellow arrow in 2012, 600 m above the terminus at the red arrow (Fig.11.10).

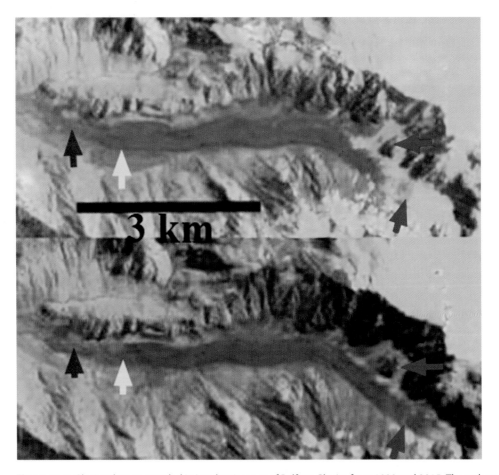

Figure 11.9 Glacier change revealed in Landsat images of Balfour Glacier from 1990 and 2015. The red arrows indicate 1990 terminus location, the yellow arrows 2015 terminus location, and the purple arrows upglacier thinning.

Figure 11.10 Stagnant debris-covered tongue of Balfour Glacier. Red arrow marks the terminus in 2012, yellow arrow the glacier river exiting the glacier, and the purple arrow the upglacier extent of limited crevassing.

11.8   Fox Glacier

Fox and Franz Josef Glaciers are the most visited glaciers in New Zealand. Fox Glacier flows west from Mount Haast and Tasman Peak. The glacier is well known for its rapid response to climate changes leading to substantial advances and retreats within limited time periods (Chinn *et al.*, 2005). At Fox Glacier advances occurred during 1985–1999 of 710 m and 2004–2008 of 290 m, with a retreat of 300 m from 1999 to 2003. Rapid retreat since 2008 has led to an overall retreat from 1990 to 2015 of 900 m (Fig. 11.11). A comparison of the annual equilibrium line altitude (ELA) and terminus behavior indicates a 3–4-year lag in the initial response, which is rapid indeed (Purdie *et al.*, 2014). The retreat since 2008 has been substantial enough to cause issues with access to the glacier by the extensive tourism industry. The retreat rate has been at a peak in 2014–2015 indicating it will continue for the next few years. Upglacier two areas of pronounced thinning are evident at the purple arrows.

11.9   Franz Josef Glacier

Franz Josef Glacier along with Fox Glacier is the most visited glacier in New Zealand (Purdie *et al.*, 2014). The glacier is well known for its rapid response to climate changes leading to substantial advances and retreats within limited time periods. The glacier has retreated over 3 km in the last 130 years, but there have been a number of advance periods during this time, with the most recent advance occurring between 1983–1999 of 1420 m and 2004–2008 of 280 m. There was a retreat from 2000 to 2003 of 400 m. Since 2008, Franz Josef Glacier has been in a period of rapid retreat, losing 800 m. Overall the glacier has retreated 950 m from 1990 to 2015, with 2014–2015 being the fastest retreat observed (Fig. 11.12).

Above the terminus on the north side of Franz Josef Glacier is the Salisbury Snowfield and Almer Glacier. In 2007 Almer Glacier reaches to within 75 m of the Franz Josef Glacier. In 2013 the terminus

Figure 11.11 Glacier change revealed in Landsat images of Fox Glacier from 1990 and 2015. The red arrows indicate 1990 terminus location, the yellow arrows 2015 terminus location, and the purple arrows upglacier thinning.

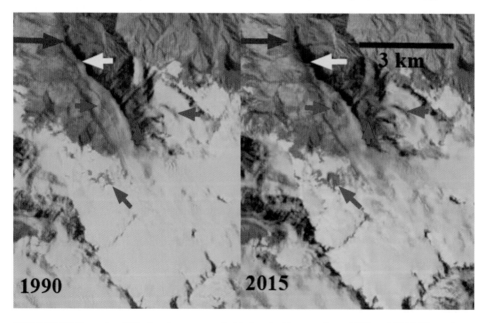

Figure 11.12 Franz Josef Glacier change revealed in Landsat images from 1990 and 2015. The red arrows indicate 1990 terminus location, the yellow arrows 2015 terminus location, and the purple arrows upglacier thinning.

is much dirtier and is 200 m from Franz Josef Glacier, with the separating distance expanding to 300 m in 2015. The icefall comparison image from 2007 and 2013 indicates the reduction in width and number of open crevasses, probably in depth too.

11.10 Classen Glacier

Classen Glacier drains east from the crest of the Southern Alps terminating in an expanding lake. The terminus is low slope and debris covered. From 1990 to 2015 the glacier has retreated 1000 m leading to expansion of the lake it ends shown in Fig. 11.13. This lake has maintained its width, suggesting the glacier is not near the limit of the basin.

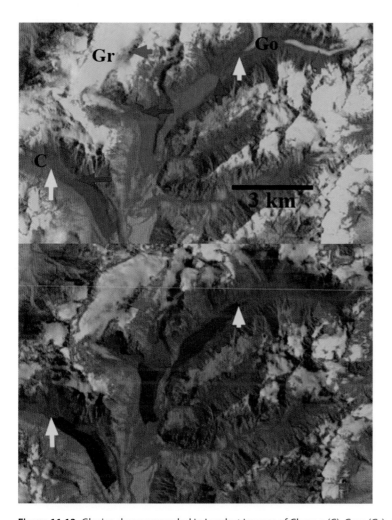

Figure 11.13 Glacier change revealed in Landsat images of Classen (C), Grey (Gr), and Godley (Go) Glaciers from 1990 and 2015. The red arrows indicate 1990 terminus location, the yellow arrows 2015 terminus location, and the purple arrows upglacier thinning.

11.11   Godley Glacier

Godley Glacier flows west from the Thumb Range terminating in a proglacial lake. The glacier is debris covered in the lowest 4 km. The expansion of debris cover is striking from 1990 to 2015, which indicates reduced flow from the accumulation zone. The glacier has retreated 1300 m with an equal amount of lake expansion.

11.12   Lyell Glacier

The Lyell and Ramsay Glaciers are the northernmost substantial valley glaciers in the Southern Alps of New Zealand (Fig. 11.14). Their combined runoff is the chief source of the Rakaia River. In 1949 Lyell glacier extended east from Rangiata Col some 7 km and Lyell Lake, which had not yet formed (Gage, 1951).

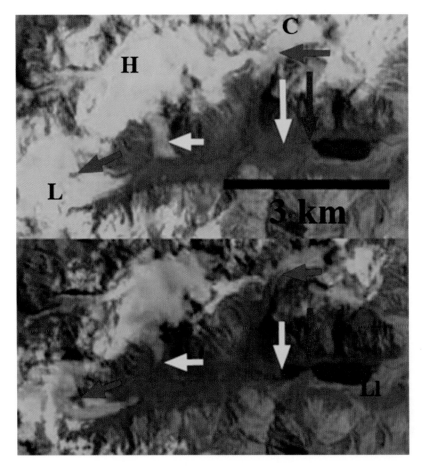

Figure 11.14 Glacier change revealed in Landsat images from 1990 and 2015 of Lyell Glacier (L) and its tributaries Cockayne (C) and Heim (H) Glaciers, which terminated at Lyell Lake (Ll) in 1990. The red arrows indicate 1990 terminus location, the yellow arrows 2015 terminus location, and the purple arrows upglacier thinning.

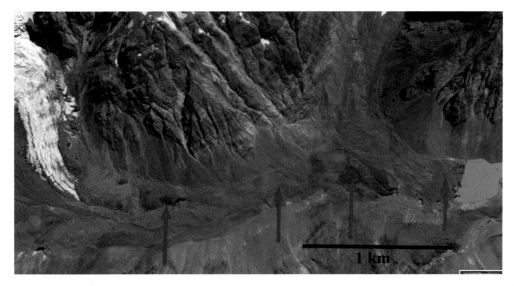

Figure 11.15 The valley below Heim and Lyell Glaciers indicating the stagnant glacier tongue of both.

In 1990 the Heim Glacier (H) reached onto the Lyell Lake valley floor (yellow arrow). In 2001 this is evident along with the fact that Lyell Lake is a single lake. The terminus of the Lyell Glacier is obscured by thick glacier cover and does end near Lyell Lake at the time; the end of the blue ice of the E tributary is not indicative of the terminus location. In 2015 Heim Glacier has retreated from the Lyell Valley and no longer is connected to the Lyell Glacier. A second small lake has formed as the terminus of Lyell Glacier has melted and retreated (red arrow). The terminus of Lyell Glacier does remain buried by debris, but it is stagnant and melting away (Fig. 11.15). Both the Lyell and Heim Glaciers have retreated 400 m from 2000 to 2015. The Lyell Glacier will likely experience a more rapid retreat in the near future as the debris-covered tongue melts away.

References

Chinn, T. (1996) New Zealand glacier responses to climate change of the past century. *New Zealand Journal of Geology and Geophysics*, **39**, 415–28.

Chinn, T. (1999) New Zealand glacier response to climate change of the past 2 decades. *Global and Planetary Change*, **22**, 155–68.

Chinn, T., Winkler, S., Salinger, J., and Haakensen, N. (2005) Recent glacier advances in Norway and New Zealand: a comparison of their glaciological and meteorological causes. *Geografiska Annaler*, **87A**, 141–57.

Dykes, R. and Brook, M. (2010) Terminus recession, proglacial lake expansion and 21st century calving retreat of Tasman Glacier, New Zealand. *New Zealand Geographer*, **66**, 203–17.

Dykes, R., Brook, M., Robertson, C., and Fuller, I. (2011) Twenty-first century calving retreat of Tasman Glacier, Southern Alps, New Zealand. *Arctic, Antarctic, and Alpine Research*, **43**, 1–10.

Gjermundsen, E., Mathieu, R., Kaab, A. *et al.* (2011) Assessment of multispectral glacier mapping methods and derivation of glacier area changes, 1978–2002, in the central Southern Alps, New Zealand, from ASTER satellite data, field survey and existing inventory data. *Journal of Glaciology*, **57** (204), 667–683. doi: 10.3189/00221431179740974

Kirkbride, M.P. (1993) The temporal significance of transitions from melting to calving termini at glaciers in the central Southern Alps of New Zealand. *The Holocene*, **3**, 232–40.

New Zealand Government (2015) Changes in glacier volume: environmental indicators, http://www.stats .govt.nz/browse_for_stats/environment/environmental-reporting-series/environmental-indicators/ Home/Atmosphere-and-climate/water-physical-stocks/glacier-volume.aspx, published 1021/2015 (accessed 4 November 2015).

Purdie, H., Anderson, B., Chinn, T. *et al.* (2014) Franz Josef and Fox Glaciers, New Zealand: historic length records. *Global and Planetary Change*, **121**, 41–52.

Quincey, D.J. and Glasser, N.F. (2009) Morphological and ice-dynamical changes on the Tasman Glacier, New Zealand, 1990–2007. *Global and Planetary Change*, **68**, 185–197.

Robertson, C., Benn, D., Brook, M. *et al.* (2012) Subaqueous calving margin morphology at Mueller, Hooker and Tasman Glaciers in *Aoraki*/Mount Cook National Park, New Zealand. *Journal of Glaciology*, **58**, 1037–1046.

Robertson, C.M., Brook, M.S., Holt, K.A. *et al.* (2013) Calving retreat and proglacial lake growth at Hooker Glacier, Southern Alps, New Zealand. *New Zealand Geographer*, **69**, 14–25. doi: 10.1111/nzg.12001

Rohl, K. (2006) Thermo-erosional notch development at fresh-water calving Tasman Glacier, New Zealand. *Journal of Glaciology*, **52** (177), 203–213. doi: 10.3189/172756506781828773

Salinger, M., and Willsman, A. (2007) Annual glacier volumes in New Zealand 1995–2005. NIWA Cline Report AKL2007-14.

Willsman, A., Chinn, T., and Lorrey, A. (2014) New Zealand Glacier Monitoring: End of summer snowline survey 2013. NIWA Technical Report. National Institute of Water and Atmosphere, Wellington.

Gage, M. (1951) The dwindling glaciers of the Upper Rakaia Valley, Canterbury, New Zealand. *Journal of Glaciology*, **1**, 504–507.

12

Alps: Mont Blanc–Matterhorn Transect

Overview

The European Alps have been the most thoroughly observed glaciated mountain range in the world. Recent inventories from Austria, France, Italy, and Switzerland reveal the same trend of accelerating mass loss and retreat since 1990. The annual terminus survey conducted on approximately 100 glaciers in Switzerland indicated that the percentage of retreating glaciers was 40% in 1985, 77% in 1995, 93% in 2005, and 89% in 2014 (VAW/ETH, 2015). In Austria the annual report of frontal change on approximately 100 glaciers was quite similar with the percentage of retreating glaciers being 32% in 1985, 86% in 1995, 92% in 2005, and 91% in 2014 (Fischer, 2015). Inventories have identified the widespread area loss. In the French Alps, Gardent *et al.* (2014) identified a 25% area loss from 1967/1971 to 2006/2009. In the Ortles–Cevedale region of the Italian Alps, Carturan *et al.* (2013) observed an area loss of 23% from 1987 to 2009. In the Swiss Alps, Fischer *et al.* (2014) found a 28% area loss from 1973 to 2010. In the Austrian Alps, Fischer *et al.* (2015) noted a 27% decline in glacier area from 1969 to 2004/2011 in the Austrian Alps. Volume losses on larger glaciers in Switzerland have been greater than on small glaciers (Paul and Haeberli, 2008).

This widespread retreat is altering the alpine landscape. In terms of water resources, the changes will be critical (Bliss, Hock and Radić, 2014). The Rhone River at Beaucaire, France, receives 25% of all August runoff from glaciers. The Rhone River is also host to hydropower plants (Pelto, 2011). The majority of power in Switzerland is produced via hydropower and in the summer glacier runoff is a key water source. Austria generates 70% of its electricity from hydropower, with many of the larger projects being glacier fed such as at Kaprun. The Kaprun power station produces 700 MW, and 60% of its water supply is from Pasterze Glacier, which is transferred through a 12 km long tunnel to Mooserboden Reservoir. The Rhone River has 19 hydropower plants supplying 25% of France's hydropower and 4% of the total energy supply. The Rhone River begins at the Rhone Glacier, Switzerland, and is fed by the largest glacier in the Alps, Gross Aletsch. The Rhone River is also fed by the glaciers of Mont Blanc and the Matterhorn region. In the Alps August runoff has increased 10–15% in many alpine basins since 1990 (Huss, 2011). Huss (2011) modeled a reduction of glacierized areas in the Alps to 12% of the current value that will lead to a sharp drop in the summer runoff contribution from glaciers, intensifying issues with water shortage in summer in many glacierized catchments. Radić *et al.* (2013) model a volume of at least 50% by 2100 for glaciers in the European Alps. We examine a pair of satellite images from 1990 and 2015 that span the Alps from Mont Blanc, France, to the Matterhorn, Switzerland, to illustrate the changes in the region. The Landsat images used are L81950282015242LGN00 and LT1950281990197XXX01 (Fig. 12.1).

Recent Climate Change Impacts on Mountain Glaciers, First Edition. Mauri Pelto.
© 2017 John Wiley & Sons, Ltd. Published 2017 by John Wiley & Sons, Ltd.

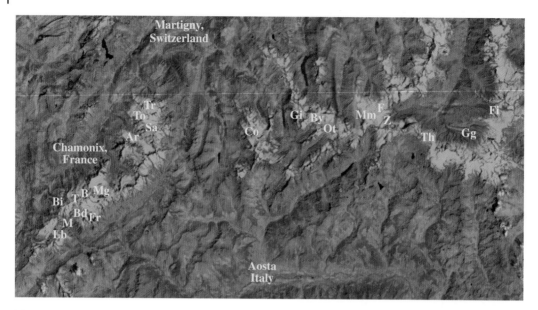

Figure 12.1 Landsat image of the glaciers in the region examined in this chapter. Bi = Bionnassay, T = Taconnaz, Mg = Mer de Glace, Ar = Argentiere, To = Tour, Tr = Trient, Sa = Saleina, Lb-Lex Blanche, M = Miage, Bd = Brouillard, Fr = Freney, Co = Corbassière, Gi = Gietro, By = Breney, Ot = Otemma, Mm = Mont Mine, F = Ferpécle, Z = Zmugg, Th = Theodulgletscher, Fi = Findelengletscher, and Gg = Gornergletscher.

12.1 Mer De Glace

Mer De Glace drains the north side of Mont Blanc. This is the largest glacier in this section of the Alps and is 12 km long (Fig. 12.2). The "sea of ice" term not only refers to the size of the glacier but also to the ogives, curved color bands formed at the base of the icefall. This sea of ice is slowing down as well as thinning and retreating. This has led to the lowest 12% of the glacier being stagnant and appearing ready to melt away in the coming decades. Vincent *et al.* (2014) model Mer de Glace into the future and generate a retreat of 1200 m by 2040; this is likely a minimum.

More people have been underneath this glacier than any other, thanks to the tunnel system that is drilled under the glacier and accessed from Montenvers Station. The glacier begins at 4200 m and ending today at 1500 m. During the Little Ice Age, the glacier advanced to the floor of the Valle de Chamonix at 1000 m.

The glacier has lost 70 m in thickness at the Montenvers Station during the last 20 years as it has retreated. This has resulted in the ongoing adjustment of the stairs leading to the tunnel system that goes under the glacier. The retreat has not been continuous as during the 1970s and 1980s, the glacier advanced 150 m. However, from 1994 to 2010 it has lost more than 500 m. As reported to the World Glacier Monitoring Service (WGMS), the retreat was 162 m from 1996 to 2000, 208 m from 2001 to 2005, and 165 m from 2006 to 2010. The terminus as shown in 2003 ends in a small proglacial lake.

The changes at the terminus between 2003 (red) and 2009 (yellow) are evident in the images shown in Fig. 12.3; the lower section of the glacier below the ice tunnel is stagnant and simply melting away. The lake that had been at the terminus in 2003 is no longer in contact with the glacier in 2009.

Below 2100 m in the relatively low slope tongue of the glacier in the ablation zone, repeat mapping indicates that the thinning rate increased dramatically from $1 \pm 0.4\,\mathrm{m\,a^{-1}}$ (years 1979–1994) to

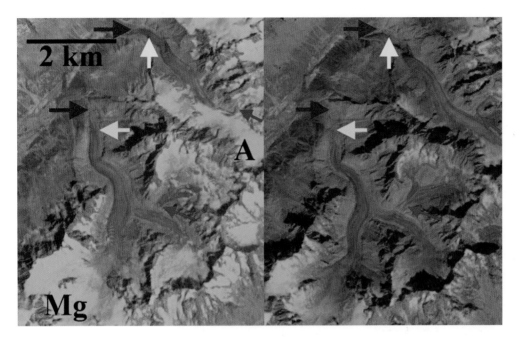

Figure 12.2 A comparison of Mer De Glace (MG) and Glacier d'Argentiere (A) in Landsat images from 1990 and 2015. The red arrows indicates the 1990 terminus, the yellow arrows the 2015 terminus, and purple arrows the upglacier thinning.

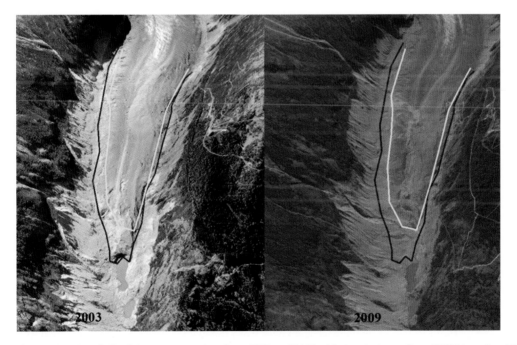

Figure 12.3 Google Earth image comparison from 2003 and 2009 with the glacier outline of 2003 in red and 2009 in yellow. (Google Earth.)

Figure 12.4 Ogives on the Mer de Glace in a 2009 Google Earth image. (Google Earth.)

$4.1 \pm 1.7 \, \mathrm{m \, a^{-1}}$ (2000–2003), according to Berthier *et al.* (2004). This is the section in the midst of the ogives, which form due to annual changes in flow through the icefall section. Also notice that the largest tributary to the Mer de Glacier has little contribution today. There are 48 identifiable ogives in the 2009 Google Earth Imagery (Fig. 12.4). The distance from the first to the last is 4.6 km, indicating an annual flow of approximately 100 m per year. Berthier and Vincent (2012) identify that 1/3 of the thinning in the lower section of the glacier is due to reduced flow from above and 1/3 to increased ablation of the glacier tongue.

12.2 Glacier d'Argentiere

Glacier d'Argentiere flows 9 km northwest toward the valley of Chamonix, France, just north of Mont Blanc and one valley north of the Mer de Glace. The glacier retreated 1000 m from 1870 to 1967. The terminus is quite crevassed indicating considerable velocity. From 1968 to 1985, the glacier advanced 300 m. Since 1985 the glacier has been retreating at an increasing rate: 80 m from 1991 to 1995, 187 m from 1996 to 2000, 199 m from 2001 to 2005, and 162 m 2006 to 2010 (WGMS, 2012). The retreat has been triggered by sustained negative mass balance. A mass balance program on Argentiere begun in 2004; the glacier lost 1.3 m in 2004, 1.9 m in 2005, 1.4 m in 2006, 0.7 m in 2007, 1.3 m in 2008, and 2.6 m in 2009. That is a cumulative mass balance loss of 9.2 m of water equivalent lost from the glacier in 6 years and a 10–11 m loss in average ice thickness from the glacier. The recent rapid retreat and mass balances losses parallel those of the other glaciers in the Alps from which data is reported to the WGMS (2013). The reason for the mass balance loss is evident from the satellite imagery of August

Figure 12.5 A 2009 Google Earth image indicating the location of the icefall that represents the active terminus. The black arrows indicate tributaries now separated, and the gray arrow indicates the one connected tributary on the north side. (Google Earth.)

2009. The snow line is at 3300 m on south-facing slopes and at 2900 meters on north-facing slopes and the main valley of the glacier; this is one month left in the melt season. At this point the glacier was 35–40% snow covered and would be less by the end of the melt season.

To be in equilibrium, a glacier must have 60% snow cover at the end of the summer. From 2004 to 2009, the average snow-covered area at the end of the melt season was 30%. The tributary glaciers draining the south-facing slopes of the valley have lost all their snow cover in 2005, 2006, and 2009. This is leading to a diminished contribution to the main stem of the Argentiere. The active terminus is at the top of a bedrock step at 2200 m, with a section of comparatively stagnant ice extending 1.2 km to the current terminus at 1720 m (Fig. 12.5). This section of the glacier is poised to rapidly melt away. Of the five main tributaries from the north, four have separated (Fig. 12.5).

12.3 Tour de Glacier

Tour du Glacier is in the Valle de Chamonix and is one valley north of Glacier d'Argentiere and two north of Mer de Glace. Here we examine the retreat of Glacier du Tour from 1990 to 2015 using Landsat imagery and Google Earth.

In 1990 the glacier terminated at 2200 m, the icefall was 1 km above the terminus, and the Aiguille du Tour tributary flowed into the Glacier du Tour. In 1999 the glacier had retreated 100 m; the Aiguille

Figure 12.6 A 2009 Google Earth image of Tour de Glacier indicating a thinning area that does not retain snow cover at Point A, the icefall region above the terminus, and the Aiguille du Tour tributary at Point C. (Google Earth.)

du Tour tributary Point C (Fig. 12.6) still reached the main glacier but is less than 200 m wide. In 2004 the terminus had retreated 200 m and the glacier is still quite crevassed near the terminus. In 2009 the area of the glacier around Point A has lost nearly all of its ice cover and is quite thin. At Point C, the Aiguille du Tour tributary has a narrow finger that reaches the main glacier. In the 2015 Landsat image, the Aiguille du Tour tributary no longer reaches the main glacier. The main terminus has retreated 500 m since 1990 and is now at the base of an icefall region just 500 m above the terminus, Point B. The activity of the icefall indicates a continued active flow. The Aiguille du Tour tributary and portion of the glacier feeding Point A do not have significant retained snow cover and are not in equilibrium.

12.4 Trient Glacier

Trient Glacier is across the Swiss border from neighboring Le Tour Glacier (Fig. 12.7). The Swiss Monitoring Network has maintained annual observations of the glacier front since 1879 (VAW/ETH, 2015). After a sustained retreat during the first half of the twentieth century, the glacier advanced 400 m from 1957 to 1987. From 1990 to 2014, the glacier has retreated 1020 m. This is faster retreat

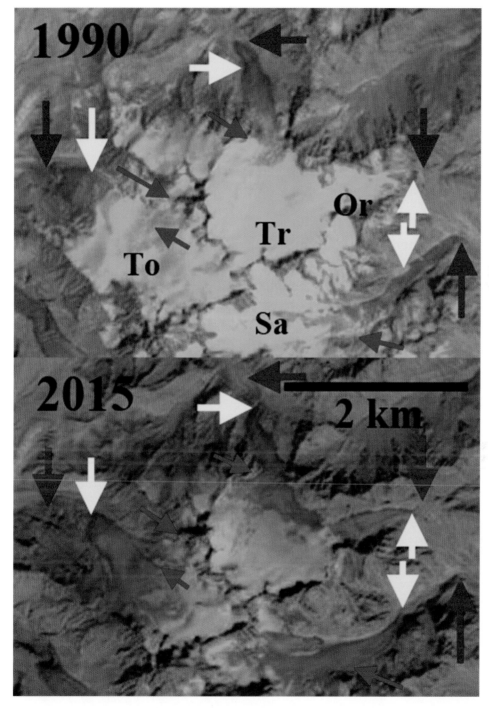

Figure 12.7 A comparison of Tour (To), Trient (Tr), Savient (Sa), and Orny (Or) Glaciers in Landsat images from 1990 and 2015. The red arrows indicate the 1990 terminus, the yellow arrows the 2015 terminus, and purple arrows the upglacier thinning.

than on the neighboring Saleina Glacier or Tour Glacier. The glacier does retain a greater area of snowpack than the adjacent glaciers, suggesting that the retreat rate will become less than the neighbors.

12.5 Saleina Glacier

Saleina Glacier is south of Trient Glacier descending a steep eastward-oriented valley from Aiguille d'Argentiere (Fig. 12.7). The Swiss Monitoring Network has maintained annual observations of the glacier front since 1878 (VAW/ETH, 2015). After a sustained retreat during the first half of the twentieth century, the glacier advanced 215 m from 1964 to 1988. From 1990 to 2014, the glacier has retreated 610 m. The lower 800 m of the glacier and the terminus are debris covered, which will slow the short-term retreat.

12.6 Bossons Glacier

Bossons Glacier flows down the west side of Mont Blanc toward Les Bossons in the Chamonix Valley. The glacier's frontal position is examined annually and reported to the WGMS. Retreat from 1990 to 1995 was 350 m. From 1995 to 2000, a slight advance occurred. From 2000 to 2010, the glacier retreated 260 m. The current active terminus is at the top of an icefall at 1700 m (Fig. 12.8). Given the active crevassing at the current terminus and an icefall 500 m further upglacier, retreat should be slow in the near future.

12.7 Taconnaz Glacier

The Taconnaz Glacier flows from 4300 m to 2000 m down the west side of Mont Blanc from the Dôme du Goûter toward the Chamonix Valley. This glacier is best known for the large avalanches that are

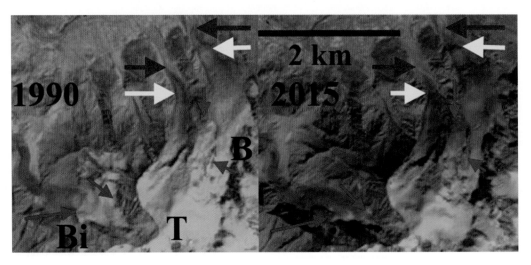

Figure 12.8 A comparison of Bossons (B), Taconnaz (T), and Bionnassay (Bi) Glaciers in a 1990 and 2015 Landsat images. Red and yellow arrows indicate the 1990 and the 2015 terminus, respectively.

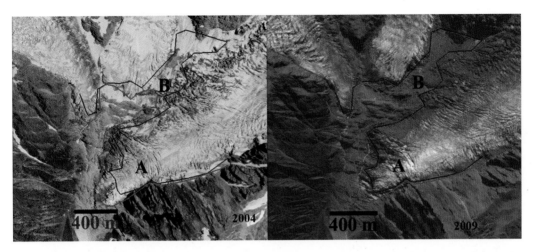

Figure 12.9 Google Earth image comparison from 2004 and 2009 indicating retreat and separation of the main glacier (A) and tributary (B) of Taconnaz Glacier. (Google Earth.)

generated by the break off of large serac ice blocks from a wide ice cliff at 3300 m during the winter (blue arrows) (Vincent *et al.*, 2004).

The ice blocks have been devastating to the inhabited areas of the Chamonix valley as recently as 1999. A comparison of 2004 and 2009 imagery in Google Earth indicates that the main terminus (A) has retreated 200 m in 5 years. The orange line is the 2004 margin and the red line is the 2009 margin (Fig. 12.9). There has also been a large retreat in the vicinity of Point B, which is where a secondary terminus used to be connected to the main valley glacier. This retreat indicates that above the terminus there is a reduction in the volume of ice heading down valley. This suggests that the retreat will continue.

The avalanche hazard has prompted construction of avalanche defenses that are well documented in photographs, such as the one in the following. This site on Mont Blanc glaciers also has nice images of the terminus glacier from 2010 indicating a much more robust terminus than Mer de Glace, based on the crevassing and glacier thickness. The avalanche protection worked in 2006 slowing an April avalanche.

12.8   Bionnassay Glacier

Bionnassay Glacier drains west from Dôme du Goûter and Aiguille de Bionnassay of the Mont Blanc Massif in France. The glacier has a heavily debris-covered terminus and has experienced less retreat from 1980 to 2010 than other Mont Blanc glaciers. Bionnassay retreated 200 m, while Mer de Glace retreated 500 m in the interval from 1998 to 2008. Bionnassay is now in rapid retreat as the stagnant terminus tongue is detached from the active glacier tongue.

In 2001 Google Earth image, the terminus is evident at the red arrow, and the regions at the green and yellow arrow are covered by glacier ice (Fig. 12.10). In 2011 the terminus has retreated 180 m since 2001, and the bedrock has emerged at the green arrow, beginning to separate the stagnant debris-covered terminus tongue. At the yellow arrow, the crevassing has diminished greatly. In 2015 the terminus has retreated to the pink arrow. Bedrock has been exposed from below the glacier

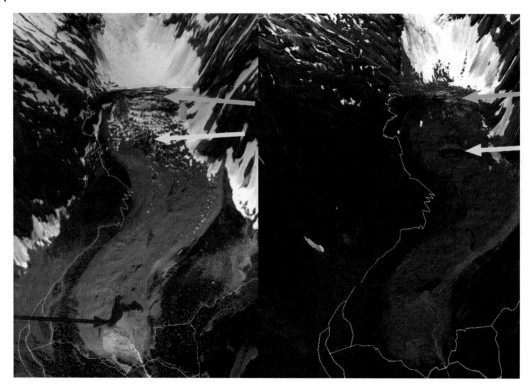

Figure 12.10 Google Earth comparison of Bionnassay Glacier in 2001 and 2011. The green and yellow arrows indicate an icefall region in 2001 that is no longer an icefall by 2011. The red arrows indicates the terminus. (Google Earth.)

terminus tongue at the yellow arrow. The active glacier terminus is now at the green arrow. The retreat of the main glacier terminus is around 200 m. However, the retreat to the newly emergent bedrock below the icefall separating the glacier is 750 m. The active terminus is now 1700 m from the 1985 terminus position at the green arrow. In the next few years, this will become a well-defined terminus, as the lower stagnant zone melt away.

12.9   Otemma Glacier

Otemma Glacier is in the Upper Rhone River watershed and feeds Lac de Mauvoisin. Climate change is altering this glacier, with terminus change not being the main story; instead it is the rising snow line and separation from tributaries. The lake fed by the glacier is impounded by Mauvoisin Dam one of the 10 largest concrete arch dams in the world. The reservoir can store 200 million cubic meters of water. The dam provides hydropower and protection against natural hazard. There are several other large glaciers in the basin Gietro, Mont Durand, and Breney that provide runoff to power what is today a large hydropower project. The Mauvoisin Dam can produce 363 MW of power.

Otemma Glacier is one of the glaciers where the terminus is monitored annually by the Swiss Glacier Monitoring Network (VAW/ETH, 2015). Here we examine the changes in this glacier from 1990 to 2015, including changes in the terminus, snow line elevation, and tributary connection during

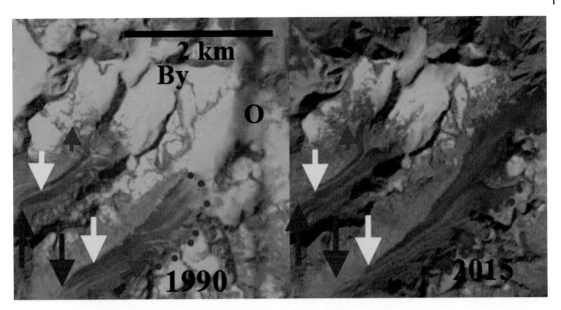

Figure 12.11 Comparison of Otemma (O) and Breney (Br) Glaciers in 1990 and 2015 Landsat images. Red and yellow arrows indicate the 1990 and the 2015 terminus positions, respectively. The snow line is indicated by blue dots on Otemma Glacier.

this period using Landsat Imagery. VAW/ETH (2015) reports that the glacier retreated at a rate of $27 \, \text{m a}^{-1}$ from 1985 to 1999 and at $40 \, \text{m a}^{-1}$ from 2000 to 2014. The total retreat from 1990 to 2015 is 900 m.

In 1988 the glacier terminates at the yellow arrow, and the snow line (green dots) extends to the divide with Bas Glacier d'Arolla at 3050 m with tributaries A, B, and C, all joining the main glacier (Fig. 12.12). In 2014 tributaries A and B are detached from the main glacier. The snow line in 2014 and 2015 almost reaches the divide with Bas Glacier d'Arolla with a few weeks left in the ablation season (Fig. 12.11 and 12.12). The area of persistent snow cover is thus restricted to the region above 3050 m. This region is not large, as the Bas Glacier d'Arolla captures most of the upper basin. That the snow line is consistently reaching the highest divide for this large glacier is noteworthy. The retreat of the large valley tongue of Otemma Glacier will remain rapid given the consistent high snow lines indicative of limited retained accumulation. Even with current climate not much of the Otemma Glacier can survive.

12.10 Breney Glacier

It is in the next valley to the north of Otemma Glacier; it also flows southwest and feeds Lac de Mauvoisin (Fig. 12.11). Breney Glacier is one of the glaciers where the terminus is monitored annually by the Swiss Glacier Monitoring Network (VAW/ETH, 2015). Here we examine the changes in this glacier from 1990 to 2015, including changes in the terminus using Landsat Imagery. VAW/ETH (2015) reports that Breney Glacier retreated 170 m from 1990 to 2000 and a further 570 m from 2000 to 2014. The total retreat from 1990 to 2015 is over 750 m. The glacier is also losing connection with the tributaries on the north side of the glacier near the purple arrow (Fig. 12.11).

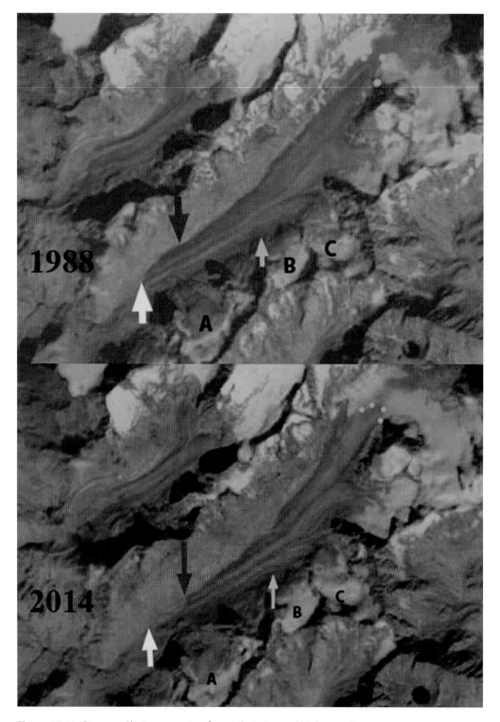

Figure 12.12 Otemma Glacier separation from tributaries and high snow line.

12.11   Gietro Glacier

Gietro Glacier may sound like a type of Italian dessert, but this glacier has a deadly history. During periods of advance, the glacier blocked off the valley of Mauvoisin. Failure of the glacier ice dam led to large flood events in 1595 and 1818 that lead to the loss of many lives in the valley below. In 1818, an advance of the Gietro glacier, now retreated high above the reservoir, generated ice avalanches that blocked the flow of the river. When the ice barrier was breached, 20 million m^3 of flood water was released devastating the valley (Collins, 1991). Erosion largely from the glaciers in the watershed produces enough sediment to cover the lake bottom with 60 cm of sediment per year (Loizeau and Dominik, 2000).

Gietro Glacier is one of the glaciers where the terminus is monitored annually by the Swiss Glacier Monitoring Network (VAW/ETH, 2015). The glacier advanced 150 m from 1962 to 1985. Retreat from 1985 to 2000 was 110 m or 7.5 m a^{-1}, accelerating to 350 m from 2004 to 2014, 35 m a^{-1}. Retreat from 1990 to 2015 has been 525 m (Fig. 12.13). In 1990 the glacier terminates at the top of a steep cliff that. In 2001 the glacier has retreated from the top of the cliff. A limited area above the terminus is stagnant – an area that is 200 m by 150 m (red arrow) (Fig. 12.13). Above this point the glacier has been thinning but still remains active as indicated by the crevassing. In 2009 the glacier is terminating on a gentler slope posing a much smaller avalanche hazard (Fig. 12.14).

The glacier in 2009, 2013, 2014, and 2015 has very limited snow cover by the end of the summer. To thrive, a glacier should be 60% snow covered at the end of the melt season; in this case, it is less than 20% – green dots mark the main snow line below. The lack of snow cover indicates a negative mass balance that is driving the retreat. If the glacier consistently loses most of the accumulation zone snow cover, it cannot survive (Pelto, 2010).

12.12   Corbassière Glacier

Corbassière Glacier is across the valley from Gietro Glacier flowing north from Grand Combin (Fig. 12.13). The runoff enters below Lac de Mauvoisin. Corbassière Glacier is one of the glaciers where the terminus is monitored annually by the Swiss Glacier Monitoring Network (VAW/ETH, 2015). The glacier advanced 80 m during the 1980s. From 1990 to 2000 the terminus retreated 110 m. From 2000 to 2014 the terminus retreat totaled 640 m, 43 m a^{-1}. Thinning upglacier has led to two tributaries on the west side of the glacier separating from the main glacier.

12.13   Glacier du Mont Miné

Glacier du Mont Miné is an 8 km long alpine glacier that drains into the Rhone River (Fig. 12.15). The runoff from this glacier is heavily tapped for hydropower; for example, Compagnie Nationale du Rhône has 19 hydropower plants on the river. Glacier du Mont Miné is one of the glaciers where the terminus is monitored annually by the Swiss Glacier Monitoring Network (VAW/ETH, 2015). The glacier experienced a period of minor advance from 1971 to 1989. From 1990 to 2000 the terminus retreated 265 m. From 2000 to 2014 the terminus retreat totaled 405 m, 28 m a^{-1}.

In the 1990 Landsat image, Glacier du Mont Miné is in an advance position in contact with the moraine at the end of a small lake that is quickly being infilled with sediment. In 1999 the glacier has retreated out of the low-lying basin below 2000 m. In 2009 the lower 300 m of the glacier is thin and

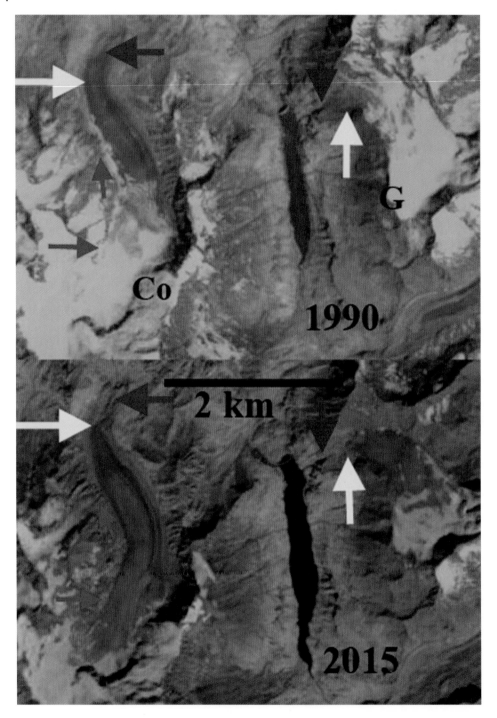

Figure 12.13 Comparison of Gietro (G) and Corbassière (Co) Glaciers in 1900 and 2015 Landsat images.

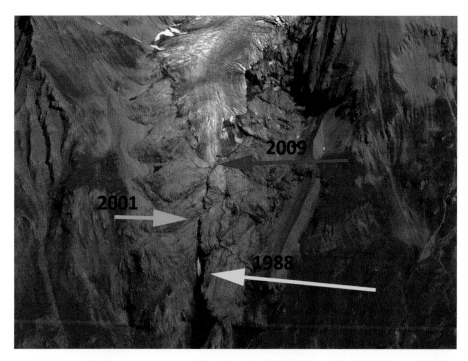

Figure 12.14 Google Earth Image of the terminus change of Gietro Glacier from 1988 to 2009. (Google Earth.)

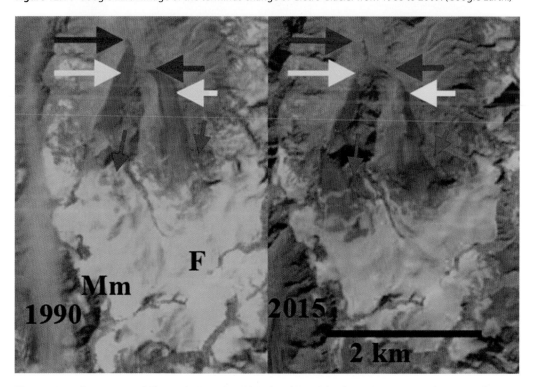

Figure 12.15 Comparison of Glacier du Mont Miné (Mm) and Ferpécle Glacier (F) in 1990 and 2015 Landsat images. Red and yellow arrows indicate the 1990 and the 2015 terminus positions, respectively. The snow line is indicated by blue dots on Otemma Glacier.

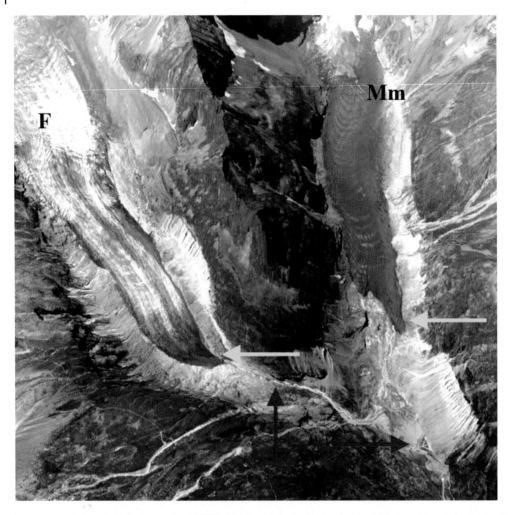

Figure 12.16 Google Earth image from 2009 of the terminus of Glacier du Mont Miné and Ferpécle Glacier. The yellow and red arrows mark the 2009 and the 1990 terminus locations, respectively. Note the lack of crevassing on the lower portion of Glacier du Mont Miné. (Google Earth.)

stagnant and has increased thin debris cover (Fig. 12.16). There is an icefall 2 km above the terminus that is now becoming dynamically detached from the glacier below. The lower terminus tongue will not survive long term without direct icefall connection.

12.14 Ferpécle Glacier

Ferpécle is just west of Glacier du Mont Miné, which was within 700 m of joining in 1990. Ferpécle Glacier is one of the glaciers where the terminus is monitored annually by the Swiss Glacier Monitoring Network (VAW/ETH, 2015). The glacier experienced a period of minor advance in the 1980s. From 1990 to 2000 the terminus retreated 130 m. From 2000 to 2014 the terminus retreat totaled 250 m, 17 m a^{-1}.

Ferpécle has retained a considerable snow-covered area above 3300 m in recent years, indicating a better mass balance, which should keep the retreat rate lower than on many Swiss glaciers.

12.15 Gornergletscher

Gornergletscher drains the northwest flank of Monte Rosa and is at the headwaters of the Vispa River. Gornergletscher is the second largest glacier in Switzerland and is one of the glaciers where the terminus is monitored annually by the Swiss Glacier Monitoring Network (VAW/ETH, 2015). From 1990 to 2000 the terminus retreated 210 m. From 2000 to 2014 the terminus retreat totaled 490 m, 35 m a^{-1}. More notable than the 600 m retreat from 1990 to 2014 of a 13.5 km long glacier is the separation of several tributary glaciers. In 1990 tributaries 2, 3, and 4 all join the main glacier. In 2015 only tributary 4 still connects (Fig. 12.17).

The lower portion of the glacier will continue to retreat rapidly as supraglacial streams and circular deflation depressions indicate a lack of flow in the lower 1.2 km of the glacier (Fig. 12.18). The glacier continues to retain a large snow-covered region above 3600 m even in warm summers. The low slope and low velocities of the glacier shortly above the terminus have led to the development of an extensive network – 20+ km of supraglacial streams and moulins (Piccini *et al.*, 2000). The moulins can remain active for 3–5 years (Piccini *et al.*, 2000) (Fig. 12.18).

12.16 Findelengletscher

Findelengletscher drains into the Vispa River in Zermatt, which joins the Rhone River. Findelengletscher is one of the glaciers where the terminus is monitored annually by the Swiss Glacier Monitoring Network (VAW/ETH, 2015). The glacier experienced a period of minor advance in the 1977–1985 period. From 1990 to 2000 the terminus retreated 400 m. From 2000 to 2014 the terminus retreat totaled 630 m, 42 m a^{-1}. The terminus is currently at 2650 m, with the lower 1.3 km having a low slope and limited crevassing, though it is not stagnant. The glacier has retained snow cover over a sizable area above 3200 m even in recent warm years. The northern tributary, Adlergletscher, below Adlerhorn has separated from the main glacier at 2950 m.

12.17 Theodulgletscher

Theodulgletscher is just west of Gornergletscher and is one of the most visited glaciers in the world (Fig. 12.18). It is part of the Zermatt ski resort in Switzerland, with three lifts crossing the glacier's western half. Since 2003 summer skiing has been enhanced with Europe's longest glacier chairlift, the 2.5 km long lift ascends from Trockener Steg, at 2940 m, to Furggsattel station at 3365 m, above the Theodulgletscher. The interesting part is that 12 of its 18 supporting masts stand on glacier ice and have been engineered so that masts can shift and revolve to accommodate the glacier's movement (Fig. 12.19).

What makes Theodulgletscher a good ski run is its modest slope and limited movement that leads to few crevasses. The ski runs themselves are well groomed and marked to keep skiers from venturing into areas where there are crevasses. With glacier retreat the distance from the glacier runs to the lift base at Trockener Steg has increased. Today an IDE snowmaker is being used to connect the

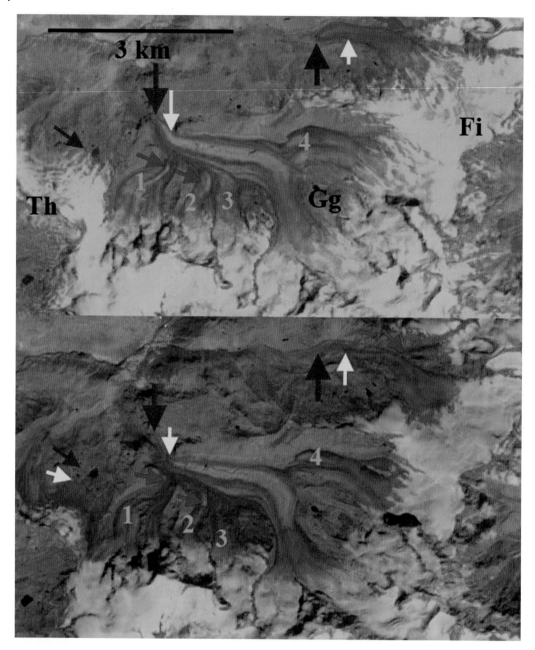

Figure 12.17 Comparison of Gornergletscher (Gg), Findelengletshcer (Fi), and Theodulgletscher (Th) in 1990 and 2015 Landsat images. The terminus positions in 1990 and 2015 are indicated by red and yellow arrows, respectively; orange numbers indicate specific Gornergletscher tributaries.

Figure 12.18 Terminus of Gornergletscher in a 2009 Google Earth image. The blue arrows indicate a supraglacial stream channel and a circular depression, and the terminus is indicated by the red arrow. (Google Earth.)

Theodulgletscher to Trockener Steg with a 700 m long strip of snow. The IDE snowmaker is the world's first snow machine that operates independently of ambient temperature.

In 1990 the terminus of the glacier reaches the lake adjacent to the ski lift base (green arrow). In 2015 the terminus has retreated 300 m from the lake.

12.18   Lex Blanche Glacier

Lex Blanche Glacier is at the southeastern edge of the Mont Blanc Massif in Italy (Fig. 12.20). The glacier advanced over 700 m from 1970 to 1990. In 1990 the glacier extended to the base of a steep slope and turned north to terminate at 1980 m. In 2015 the glacier has retreated 1100 m and terminates at 2450 m and remains on a steep slope. The glacier is heavily crevassed a short distance above the terminus, suggesting the period of rapid retreat should be ending. A tributary from the north has detached from the main glacier at the purple arrow.

Figure 12.19 Theodulgletscher in a 2009 Google Earth image. Pink arrow indicates bedrock knob emerging in the middle of the glacier. (Google Earth.)

12.19 Miage Glacier

Miage Glacier is a heavily debris-covered glacier that reaches the floor of Val Veny at 1730 m. The glacier thickened in the terminus reaching during the 1970–1990 period but did not advance significantly. Since 1990 the terminus tongue has been downwasting but not retreating. The glacier is developing an increasing number and size of supraglacial lakes, much like on Himalayan glaciers. A period of rapid retreat will occur as the lakes merge and the ice under the debris continues to melt (Fig. 12.21). The debris cover thickness measured ranged from 0.04 to 0.55 m with melt rates averaging $14–16 \ \mathrm{mm \ d^{-1}}$ compared to $46–58 \ \mathrm{mm \ d^{-1}}$ on clean glacier ice (Brock *et al.*, 2010). Each of the glaciers that drain into the Val Veny supply the Dora Baltea River that has 41 MW of installed hydropower.

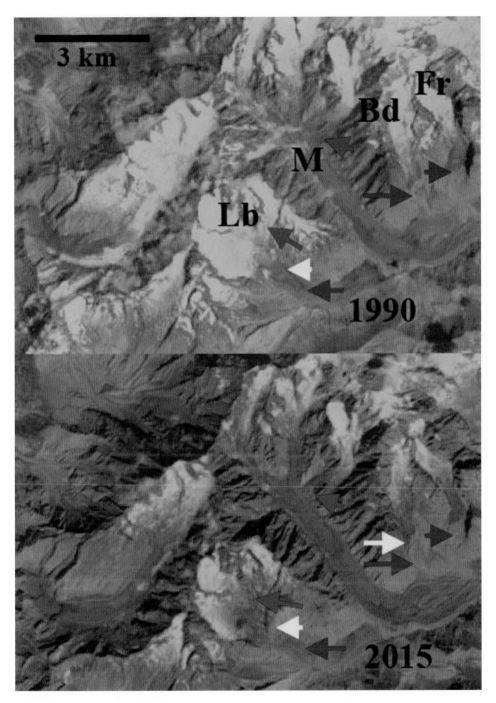

Figure 12.20 Comparison of Lex Blanche (Lb), Miage (M), Brouillard (Br), and Freney (Fr) Glaciers in 1990 and 2015 Landsat images. The terminus positions in 1990 and 2015 are indicated by red and yellow arrows, respectively.

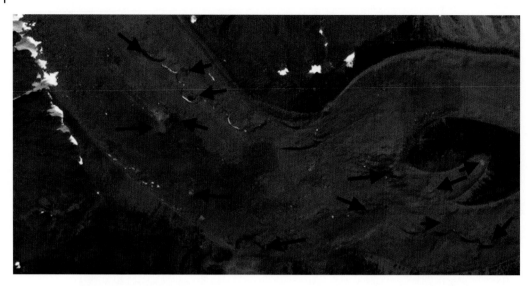

Figure 12.21 Miage Glacier debris-covered terminus region. Black arrows indicate supraglacial lakes in this 2011 Google Earth image. (Google Earth.)

12.20 Brouillard Glacier

Brouillard Glacier flows east from Mont Blanc starting at 3800 m and terminating on a steep slope above the Val Veny. In 1990 the terminus was at 2200 m. In 2015 the glacier had retreated 550 m to terminate at 2450 m. The glacier is heavily crevassed close to the terminus suggesting the retreat will be slow.

12.21 Freney Glacier

Freney Glacier flows east from Mont Blanc starting at 3800 m and terminating on a steep slope above the Val Veny just north of Brouillard Glacier. The glacier terminated at 2200 m in 1990. In 2015 the glacier had retreated 400–2420 m. The glacier is heavily crevassed close to the terminus suggesting the retreat will be slow.

References

Berthier, E., Arnaud, Y., Baratoux, D. *et al.* (2004) Recent rapid thinning of the 'Mer de Glace' glacier derived from satellite optical images. *Geophysical Research Letters*, **31** (17), L17401. doi: 10.1029/2004GL020706

Berthier, E. and Vincent, C. (2012) Relative contribution of surface massbalance and ice-flux changes to the accelerated thinning of Mer de Glace, French Alps, over 1979–2008, *J. Glaciol.*, **58**, 501–512, doi: 10.3189/2012JoG11J083

Bliss, A., Hock, R., and Radić, V. (2014) Global response of glacier runoff to twenty-first century climate change. *J. Geophys. Res.: Earth Surf.*, **119** (4), 717–730. doi: 10.1002/2013JF002931

Brock, B. W., Mihalcea, C., Kirkbride, M., Diolaiuti, G., Cutler, M. and Smiraglia, C. (2010) Meteorology and surface energy fluxes in the 2005–2007 ablation seasons at the Miage debris-covered glacier, Mont Blanc Massif, Italian Alps, *J. Geophys. Res.*, **115**, D09106. doi: 10.1029/2009JD013224

Carturan, L., Filippi, R., Seppi, R. *et al.* (2013) Area and volume loss of the glaciers in the Ortles-Cevedale group (Eastern Italian Alps): controls and imbalance of the remaining glaciers. *The Cryosphere*, 7, 267–319.

Collins, D.N. (1991) Climatic and glaciological influences on suspended sediment transport from an alpine glacier, in *Sediment and Stream Water Quality – Proceedings of a Symposium*, vol. **203**, IAHS, Vienna, pp. 3–12.

Fischer, A. (2015) Sammelbericht über die Gletschermessungen des Österreichischen Alpenvereins im Jahre 2014. *Letzter Bericht: Bergauf 02/2014*, **Jg. 69** (139), 34–40. https://www.alpenverein.at/portal_wAssets/docs/museum-kultur/Gletschermessdienst/Gletscherbericht-2013_2014-_-Bergauf-02-2015.pdf

Fischer, M., Huss, M., Barboux, C., and Hoelzle, M. (2014) The new Swiss Glacier Inventory SGI2010: Relevance of using high resolution source data in areas dominated by very small glaciers. *Arctic, Antarctic, and Alpine Research*, **46**, 935–947. doi: 10.1657/1938-4246-46.4.933

Fischer, A., Seiser, B., Stocker Waldhuber, M. *et al.* (2015) Tracing glacier changes in Austria from the Little Ice Age to the present using a lidar-based high-resolution glacier inventory in Austria. *The Cryosphere*, **9**, 753–766. doi: 10.5194/tc-9-753-2015

Gardent, M., Rabatel, A., Dedieu, J.-P., and Deline, P. (2014) Multitemporal glacier inventory of the French Alps from the late 1960s to the late 2000s. *Global Planet. Change*, **120**, 24–37. doi: 10.1016/j.gloplacha.2014.05.004

Huss, M. (2011), Present and future contribution of glacier storage change to runoff from macroscale drainage basins in Europe, *Water Resour. Res.*, **47**, W07511. doi: 10.1029/2010WR010299 DOI: 10.1029/2010WR010299#_blank#Link to external resource: 10.1029/2010WR010299 .

Loizeau, J.-L. and Dominik, J. (2000) Evolution of the Upper Rhone River discharge and suspended sediment load during the last 80 years and some implications for Lake Geneva. *Aquatic Sciences*, **62**, 57–64.

Paul, F. and Haeberli, W. (2008) Spatial variability of glacier elevation changes in the Swiss Alps obtained from two digital elevation models. *Geophysical Research Letters*, **35** (21), L21502. doi: 10.1029/2008GL034718

Pelto, M. (2010) Forecasting temperate alpine glacier survival from accumulation zone observations. *The Cyrosphere*, **3**, 323–350.

Pelto, M. (2011) Hydropower: hydroelectric power generation from alpine glacier melt, in *Encyclopedia of Snow, Ice and Glaciers* (eds V. Singh, P. Singh, and U. Haritashya), Springer, Netherlands, pp. 546–551. doi: 10.1007/978-90-481-2642-2_624

Piccini, L. *et al.* (2000) Moulins and marginal contact caves in the Gornergletscher, Switzerland. *Nimbus*, **23–24**, 94–99.

Radić, V., Bliss, A., Beedlow, A.C. *et al.* (2013) Regional and global projections of twenty-first century glacier mass changes in response to climate scenarios from global climate models. *Climate Dynamics*, **42** (1–2), 37–58. doi: 10.1007/s00382-013-1719-7

VAW/ETH (2015) *Swiss Glacier Monitoring Network: Glacier Length List*, http://glaciology.ethz.ch/swiss-glaciers/ (last accessed 28 May 2016).

Vincent, C., Harter, M., Gilbert, A. *et al.* (2014) Future fluctuations of Mer de Glace, French Alps, assessed using a parameterized model calibrated with past thickness changes. *Annals of Glaciology*, **55** (66), 15–24.

Vincent, C., Kappenberger, G., Valla, F. *et al.* (2004) Ice ablation as evidence of climate change in the Alps over the 20th century. *Journal of Geophysical Research*, **109** (D10), D10104. doi: 10.1029/2003JD003857

WGMS (2012) *Fluctuations of Glaciers 2005–2010 (Vol. X).* ICSU(WDS)/IUGG(IACS)/UNEP/UNESCO/WMO, World Glacier Monitoring Service, Zurich. doi: 10.5904/wgms-fog-2012-11

WGMS (2013) *Glacier Mass Balance Bulletin No. 12 (2010–2011).* ICSU(WDS)/IUGG(IACS)/UNEP/UNESCO/WMO, World Glacier Monitoring Service, Zurich. doi: 10.5904/wgms-fog-2013-11

13

Alpine Glacier Change Summary

Loss of area and volume of alpine glaciers is persistent across alpine ranges of the world. The homogenous response of glaciers from region to region indicates the global nature of the climate change when forcing a glacier response (Zemp *et al.*, 2015). The retreat of 162 of the 165 glaciers described in the volume quantifies the response. The volume loss that has occurred led to 1.1 mm a^{-1} sea level rise during the twentieth century (Marzeion, Jarosch, and Hofer, 2012). The volume of water in these glaciers remaining can provide an additional 1.2–2.1 mm a^{-1} of sea level rise during the twenty-first century (Marzeion, Jarosch, and Hofer, 2012; Radic and Hock, 2011). Geographically glacier retreat is leading to the development and expansion of numerous alpine lakes. Regionally glaciers act as natural reservoirs, storing water in a frozen state both seasonally and for longer periods. Glaciers modify streamflow, releasing the most runoff during the warmest period of the year, which in most cases is the driest period when all other sources of water are at a minimum (Fountain and Tangborn, 1985). The exception is the Himalayas where the melt season is also the monsoon season. This seasonal variation characteristic mitigates low flow intervals and makes glacier runoff a valuable water resource for hydropower, agriculture, and aquatic life (Bliss, Hock, and Radić, 2014; Nolin *et al.*, 2010; Pelto, 2011). The amount of glacier runoff is significant not just in rivers but also for certain marine regions. Neal, Hood, and Smikrud (2010) identified that 47% of the annual freshwater runoff to the Gulf of Alaska is from glaciers. Salmon populations in the Nooksack River that flows into Puget Sound region are sensitive to glacier runoff particularly during late summer and fall runs (Pelto, 2015). Glacier retreat is also generating more glacier moraine-dammed lakes that are prone to failure such as at Kedarnath, India (Das, Kar, and Bandyopadhyay, 2015) (Fig. 13.1).

The amount of runoff provided by a glacier is the product of its surface area and ablation rate (Pelto, 2008). Glacier runoff does not increase or decrease the long-term runoff for a basin; total runoff over a period of several years is determined largely by annual precipitation. Modeling by Huss *et al.* (2014) indicates that the rapid glacier wastage and a transient runoff increase are followed by reduced melt season discharge as the glacier area declines. Glacier volume loss contributes to changes in streamflow, leading to an increase in overall streamflow if the rate of volume loss is sufficiently large (Stahl and Moore, 2006) or a decline in streamflow if the area of glacier cover declines sufficiently to offset any increase in ablation rate (Huss *et al.*, 2014). Observations and modeling studies in numerous glaciated mountain ranges indicate that a climate warming induced shift toward a more nival-based runoff signal with an earlier peak runoff and eventually a decline in glacier runoff and late summer river discharge as glacier area and volume are reduced is underway (Pellicotti *et al.*, 2011; Huss *et al.*, 2014). In some mountain areas, glacier runoff has already peaked in specific watersheds, while in other watersheds, the peak may not occur until 2100 (Pelto, 2008; Sorg *et al.*, 2012; Huss *et al.*, 2014).

The continued increase in mean annual mass losses indicates that glacier retreat will continue and that retreat to date has not brought alpine glaciers in any of the 10 regions examined close to an

Recent Climate Change Impacts on Mountain Glaciers, First Edition. Mauri Pelto.
© 2017 John Wiley & Sons, Ltd. Published 2017 by John Wiley & Sons, Ltd.

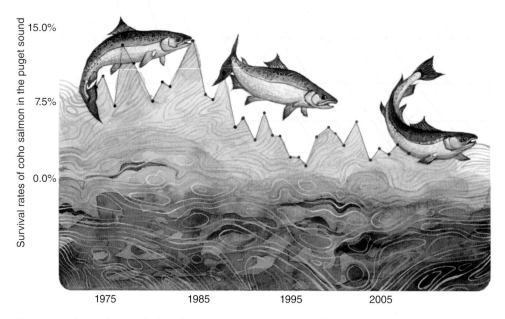

Figure 13.1 Survival rates of coho salmon in Puget Sound, surveyed by the Salish Sea Marine Survival Project (Zimmerman *et al.*, 2015). (Courtesy of Jill Pelto.)

equilibrium state. In most regions mass loss, area loss, and retreat have increased since the start of the twenty-first century (Zemp *et al.*, 2015).

References

Bliss, A., Hock, R., and Radić, V. (2014) Global response of glacier runoff to twenty-first century climate change. *Journal of Geophysical Research: Earth Surface*, **119** (4), 717–730. doi: 10.1002/2013JF002931

Das, S., Kar, N.S., and Bandyopadhyay, S. (2015) Glacial lake outburst flood at Kedarnath, Indian Himalaya: a study using digital elevation models and satellite images. *Natural Hazards*, **77** (2), 769–786.

Fountain, A. and Tangborn, W. (1985) The effect of glaciers on streamflow variations. *Water Resources Research*, **21**, 579–586.

Huss, M., Zemp, M., Joerg, P.C., and Salzmann, N. (2014) High uncertainty in 21st century runoff projections from glacierized basins. *Journal of Hydrology*, **510**, 35–48.

Marzeion, B., Jarosch, A., and Hofer, M. (2012) Past and future sea-level change from the surface mass balance of glaciers. *The Cryosphere*, **6**, 1295–1322. doi: 10.5194/tc-6-1295-2012

Neal, E.G., Hood, E., and Smikrud, K. (2010) Contribution of glacier runoff to freshwater discharge into the Gulf of Alaska. *Geophysical Research Letters*, **37**, L06404. doi: 10.1029/2010GL042385

Nolin, A., Phillippe, J., Jefferson, A., and Lewis, S. (2010) Present-day and future contributions of glacier runoff to summertime flows in a Pacific Northwest watershed: implications for water resources. *Water Resources Research*, **46** (12), W12509.

Pellicotti, F., Basuder, A., and Parola, M. (2011) Effect of glaciers on streamflow trends in the Swiss Alps. *Water Resources Research*, **46**, W10522. doi: 10.1029/2009WR009039

Pelto, M. (2008) Impact of climate change on North Cascade alpine glaciers and alpine runoff. *Northwest Science*, **82** (1), 65–75.

Pelto, M. (2011) Hydropower: hydroelectric power generation from alpine glacier melt, in *Encyclopedia of Snow, Ice and Glaciers* (eds V. Singh, P. Singh, and U. Haritashya), Springer, Netherlands, pp. 546–551. doi: 10.1007/978-90-481-2642-2_624

Pelto, M. (2015) Climate driven retreat of mount baker glaciers and changing water resources, in *SpringerBriefs in Climate Studies*, Springer International Publishing. doi: 10.1007/978-3-319-22605-7

Radic, V. and Hock, R. (2011) Regionally differentiated contribution of mountain glaciers and ice caps to future sea-level rise. *Nature Geoscience*, **4**, 91–94.

Sorg, A., Bolch, T., Stoffel, M. *et al.* (2012) Climate change impacts on glaciers and runoff in Tien Shan (Central Asia). *Nature Climate Change*, **2**, 725–731.

Stahl, K. and Moore, R.D. (2006) Influence of watershed glacier coverage on summer streamflow in British Columbia, Canada. *Water Resources Research*, **42**. doi: 10.1029/2006WR005022

Zemp and others (2015) Historically unprecedented global glacier decline in the early 21st century. *Journal of Glaciology*, **61** (228), 745–763. doi: 10.3189/2015JoG15J017

Zimmerman, M., Irvine, J., O'Neill, M. *et al.* (2015) Spatial and Temporal Patterns in Smolt Survival of Wild and Hatchery Coho Salmon in the Salish Sea. *Marine and Coastal Fisheries: Dynamics, Management, and Ecosystem Science*, **7** (1), 116–134. doi: 10.1080/19425120.2015.1012246

Index

Recent Climate Change Impacts on Mountain Glaciers, First Edition. Mauri Pelto.
© 2017 John Wiley & Sons, Ltd. Published 2017 by John Wiley & Sons, Ltd.